Museum Field Columbian

The birds of eastern North America known to occur east of the nineteenth meridian

Museum Field Columbian

The birds of eastern North America known to occur east of the nineteenth meridian

ISBN/EAN: 9783337221836

Printed in Europe, USA, Canada, Australia, Japan

Cover: Foto ©berggeist007 / pixelio.de

More available books at **www.hansebooks.com**

EASTERN NORTH AMERICA

WATER BIRDS

PART I

KEY TO THE FAMILIES AND SPECIES

BY

CHARLES B. CORY

Curator of Department of Ornithology in the Field Columbian Museum

SPECIAL EDITION

PRINTED FOR THE

FIELD COLUMBIAN MUSEUM

CHICAGO, ILL.

1899

ARTIFICIAL KEY

TO DISTINGUISH THE

LAND BIRDS AND WATER BIRDS.

A

All birds having **toes more or less webbed and no feathers on the tarsus** (all intergradations between figures A and B), except the Vultures,* and also all birds having the **bill more than three inches long**, whether the toes are webbed or not, are **Water Birds.** There are also a few Water Birds which do not agree with the above description, which may be described as follows:—

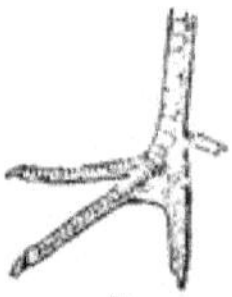

B

Shore Birds.

Gallinules.

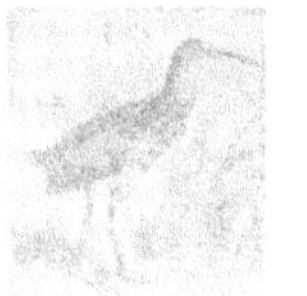

Rails.

Least Bitterns and Small Herons.

SHORE BIRDS. — The majority of the Shore Birds either have the toes with small web, or the bill over three inches long. Those which do not may be recognized by the following characters: hind toe, when present, elevated above level of front toes; bill, rather soft; nostril, a slit; lower portion of tibia (upper part of leg), not feathered: **first primary, about equal to second.**

GALLINULES. — Rail-like birds, inhabiting marshy places; forehead, covered by a horny plate or shield.

RAILS. — Wing, rather short and rounded; toes, long: hind toe, on level with front toes (true Rails); **first primary, much shorter than second.**

LEAST BITTERNS AND SMALL HERONS. — Toes, four; lores, bare; inner side of middle toe nail, with comb-like edge.

All other birds are called Land Birds.

* Several **Land Birds** have the toes partly webbed, such as the Goatsuckers, *Caprimulgidae*, the Vultures, *Cathartidae*, and a few others, but all have the tarsus more or less feathered, except the Vultures, the only Land Birds which have webbed toes, and no feathers on the tarsus.

THE BIRDS

OF

EASTERN NORTH AMERICA

KNOWN TO OCCUR EAST OF THE NINETIETH MERIDIAN

WATER BIRDS

PART I

KEY TO THE FAMILIES AND SPECIES

BY

CHARLES B. CORY

Curator of the Department of Ornithology in the Field Columbian Museum, Chicago; Vice-President of the American Ornithologists' Union; Fellow of the Linnaean and Zoological Societies of London; Member of the British Ornithologists' Union; Honorary Member of the California Academy of Sciences; Cor. Member New York Academy of Sciences, etc., etc.

AUTHOR OF "THE BEAUTIFUL AND CURIOUS BIRDS OF THE WORLD," "THE BIRDS OF THE BAHAMA ISLANDS," "THE BIRDS OF HAITI AND SAN DOMINGO," "THE BIRDS OF THE WEST INDIES," "A NATURALIST IN THE MAGDALEN ISLANDS," "HUNTING AND FISHING IN FLORIDA," "KEY TO THE WATER BIRDS OF FLORIDA," "HOW TO KNOW THE SHORE BIRDS OF NORTH AMERICA," "HOW TO KNOW THE DUCKS, GEESE, AND SWANS," ETC., ETC.

SPECIAL EDITION PRINTED FOR THE

FIELD COLUMBIAN MUSEUM, CHICAGO, ILL.

1899

ALFRED MUDGE & SON, PRINTERS,
24 FRANKLIN STREET.

THE PAGES IN THIS VOLUME HAVE BEEN INTERLEAVED WITH AN ACID FREE PAPER TO PERMIT BINDING AND TO REDUCE FURTHER DETERIORATION.

PREFACE

Ornithology is the science of birds (Gr., *ὄρνιθος*, *ornithos*, of a bird; *λόγος*, *logos*, a discourse), and to become an expert ornithologist requires years of hard work, combined with a love for the study itself; but there are many students of nature who would like to know the birds about them, but do not have the time nor desire to go deeply into the subject. To meet such a want, the present "keys" have been prepared, in which the *species are grouped according to size*, and it is believed they will enable the novice to accurately identify any of our birds.

Careful comparison of large series of birds has shown that while adult birds of the same species differ considerably in length, the wing measure is very constant, the variation in a large number of specimens being so small that, allowing for possible extremes, we may safely arrange our birds in groups according to length of wing. A Song Sparrow may vary slightly in size, but the largest Song Sparrow is never as large as the smallest adult Robin, and *vice versa*. By grouping the various birds from the Humming Bird to the Eagle and Albatross, according to the length of wing (allowing, of course, for unusual extremes), the identification of any species then becomes a comparatively simple matter, as usually the birds contained in each group are so few in number that characteristic differences in each species are easily indicated.

Let us, for example, assume we have a bird before us which we wish to identify; we first should find out to which family it belongs. Turning to the Key to Families, page 10, we find this an easy matter (as the families are few and the illustrated differences in the bills and feet very characteristic), we discover our bird to be a duck. Having ascertained the family to which our bird belongs, we turn to the Key to Species. We have, of course, measured the wing and found it to be 5.90 inches long, measured from the carpus (bend of wing) to tip. (See illustration, "How to Measure a Bird," on page 8.)

We now turn to the Ducks, and discover they are divided into subfamilies, the Bay and Sea Ducks having a flap or lobe on the hind toe, and the fresh-water ducks, or River and Pond Ducks, have no large flap on the hind toe. Our bird has a flap on its hind toe, and is evidently a salt-water duck, belonging to the subfamily *Fuligulinæ*. This subfamily, we find, contains two sections. Section I having the tail feathers *not* stiff and pointed, and Section 2, tail feathers stiff and pointed. Our duck has stiff, pointed tail feathers, and therefore belongs in Section 2. Section 2 contains two species; one having whole front of head and cheeks *black*, the other, with sides of head more or less *white*. As our duck has a patch of white on the side of the head, it must, therefore, be the Ruddy Duck, *Erismatura jamaicensis*.

All measurements of birds are given in inches and fractions of an inch. The diagrams on page 8 will illustrate how a bird should be measured, and the chart (pages viii and ix) will be useful to the young student of ornithology who may not be familiar with the technical terms used in describing birds. Such terms as primaries and axillars should be learned at once. It is customary to indicate the sexes by the signs of Mars and Venus; the male, of course, being given that of Mars, ♂, and the female, Venus, ♀.

In preparing the Key, a very large number of birds were examined and measured. In addition to the large collection of North American birds contained in the Field Columbian Museum, by courtesy of my friends, Dr. J. A. Allen and Prof. Robert Ridgway, the collections of the Smithsonian Institution and the American Museum of Natural History were always open to me, as well as the splendid private collection of Mr. William Brewster, at Cambridge, Mass.

A few species which occur in Greenland, but which have not been observed elsewhere in North America, and rare exotic stragglers have been excluded from the Key, but are given in their proper place in the body of the work.

The larger portion of the illustrations are original and are the work of Mr. Edward Knobel, of Boston. Numerous cuts are also included by arrangement with Messrs. Little, Brown & Co., of Boston, which are taken from Baird, Brewer & Ridgway's North American Birds. A few others were obtained from Messrs. Estes & Lauriat, used in Dr. Elliott Coues' Key to North American Birds.

C. B. CORY.

CONTENTS.

GLOSSARY.

Nearly all the terms used in describing a bird may be more easily and clearly understood by examining the illustrations of "bird topography," on pages viii and ix, than from a written description; a few, however, may require a word of explanation.

Cere. — A hard skin-like covering on the base of the upper mandible (Parrots, Hawks, etc.).

Mandibles. — Some authors use the word *maxilla* for the upper half of the bill, and *mandible* for the lower. I prefer, however, to describe the two halves of the bill as upper and lower mandible.

Culmen. — The ridge of the upper mandible.

Gonys. — Lower outline (middle) of under mandible.

Unguis. — The nail on the end of the upper mandible; very pronounced in several families of water birds (Ducks, Pelicans, and Petrels).

Axillars or Axillary Plumes. — Several elongated feathers at the junction of the wing and body. (Lat. *axilla*, the arm-pit.)

Carpus or Carpal Joint. — Bend of the wing. The third segment of the wing corresponding to the wrist (see illustration).

Speculum. — A wing band or patch (usually of a different color from the rest of the wing), formed by the terminal portion of the secondaries; very noticeable in the Ducks.

Tarsus. — Extends from the root of the toes to the end of the tibia (what *appears* to be the bend of the leg or knee; but which is, in reality, the heel joint). See illustration.

Toes Syndactyle. — Outer and middle toes more or less joined together (Kingfisher, etc.).

Toes Zygodactyle. — Arranged in pairs, two in front, two behind (Cuckoos, etc.).

Tarsus Reticulate. — Covered with numerous small, uneven scales (Plovers, etc.).

Tarsus Scutellate. — Comparatively large, somewhat squarescut scales, one above the other, covering the front of the tarsus.

Booted Tarsus. — Scales fused together on greater portion of tarsus so as to be indistinct or invisible except on lower part (Robins, etc.).

Superciliary Stripe. — Stripe over the eye.

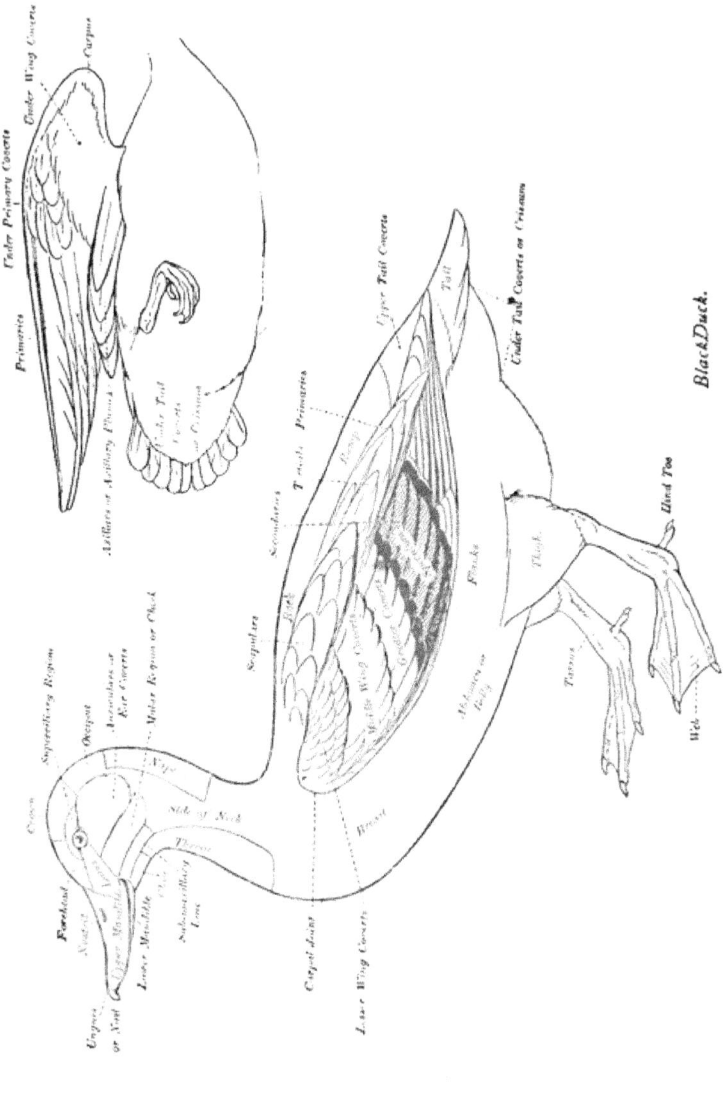
Under Primary Coverts
Under Wing Coverts
Carpus
Primaries
Axillars or Axillary Plumes
Under Tail Coverts or Crissum
Upper Tail Coverts
Tail
Under Tail Coverts or Crissum
Black Duck.
Primaries
Tertials
Rump
Secondaries
Hind Toe
Thigh
Flanks
Back
Scapulars
Greater Coverts
Middle Wing Coverts
Abdomen or Belly
Tarsus
Web
Superciliary Region
Occiput
Auriculars or Ear Coverts
Malar Region or Cheek
Nape
Crown
Side of Neck
Throat
Breast
Forehead
Nostril
Upper Mandible
Lower Mandible
Submaxillary Line
Carpal Joint
Lesser Wing Coverts
Unguis or Nail

TOPOGRAPHY OF A DUCK

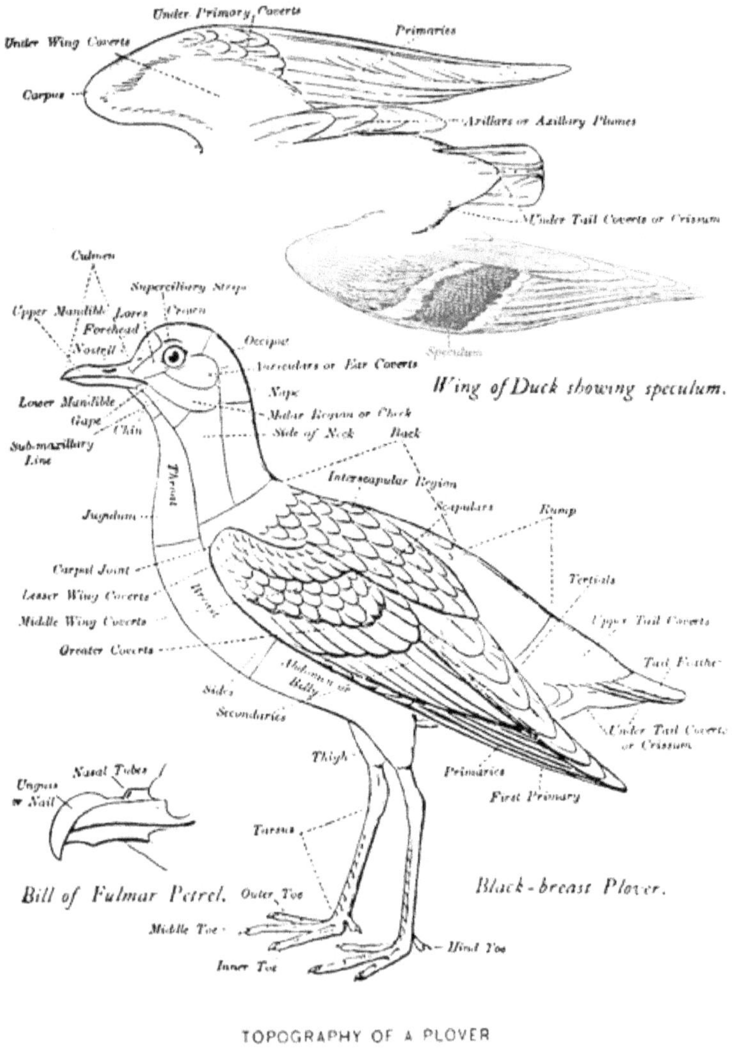

TOPOGRAPHY OF A PLOVER

INTRODUCTION.

A BIRD

Is a feathered vertebrate animal; or, to describe it more fully, it is an air-breathing, warm-blooded, feathered, oviparous (egg-laying), vertebrate animal, having a four-chambered heart, and a complete double circulation. Birds occupy a place in nature intermediate between the mammals and the reptiles, and many naturalists consider a bird to be merely a modified reptile. Be that as it may; we are morally certain that thousands of years ago there existed on the earth huge, lizard-like birds, of many of which we know nothing. The oldest known form of which we have any actual knowledge is the celebrated *Archæopteryx*, a fossil found by Andreas Wagner, in the Oölitic slate of Solenhofer, Bavaria, in 1861. This reptile bird had a lizard-like tail bordered with feathers, and jaws armed with teeth.

Of late years, many important osteological discoveries have been made, and from

Dodo, Ostrich. Moa.

the reconstructed skeletons we are able to form some idea of the size and shape of a few of the many huge and strange birds which lived and died in the forgotten past. The *Harpagornis*, an immense raptorial bird, or some similar monster, may have originated the stories of the Roc of nursery lore. Still later, we have the Dodo of Mauritius, and the Moa of New Zealand, the latter a giant bird, much larger than the largest Ostrich, which, it is claimed, was still in existence at the beginning of the present century, and a few of the older natives claim to have heard their fathers talk of seeing it alive. We know of several species which have become extinct during the past fifty years, notably the Great Auk and the Labrador Duck, *Camptolæmus labradorius*.

In the West Indies we have numerous instances of the recent disappearance of insular forms. The Jamaica Petrel, *Aestrelata caribbæa*, is now supposed to have been exterminated by the Mongoose (*Herpestes*), which animal was introduced into the island some years ago for the purpose of killing the rats. Wallace, in his "Distribution of Animals," states that no less than six species of Parrots were said to have been formerly found in Guadaloupe and Martinique, but at the present time no Parrots are known to exist on either of those islands.

Ledru, in his "Voyage aux iles Teneriffe," published in Paris, in 1810, mentions several birds which are now unknown, among them a green pigeon from St. Thomas.

While many species have become extinct, others have extended their range, and, accommodating themselves to changed conditions and environment, have in the course of time developed new forms. Birds showing decided and constant differences are recognized as *species*, whereas if the differences in color or size are not very great and intermediate forms occur showing an intergradation from one to the other, they are called *races* or *subspecies*. Races are really species in process of development and are caused by difference in climate, food, etc. Insular forms which are sufficiently removed from the parent stock to warrant the belief that their isolation is complete, are generally recognized as species even though the differences would hardly be worthy of specific recognition if the two forms occurred in close proximity on the main land. It should be borne in mind, however, that the line of demarkation between a species and a subspecies is a purely arbitrary one and is largely a matter of individual opinion. Subspecies are distinguished by a third name; for example, *Dendroica palmarum hypochrysea* is a race or subspecies of *Dendroica palmarum*.

Having learned something as to what a bird is, let us take up in order the more important external parts, such as the wing, tail, bill, and feet.

THE WING.

As an aid to identification, the wing characters are most important. The terms **primaries, secondaries, axillars, wing coverts,** etc., are constantly used in describing birds, and the student should learn to recognize them at a glance.

The Remiges are the flight feathers of the wing, and the *Tectrices* are the small feathers covering the upper part of the wing or shoulder (see illustration), and are usually called **coverts.** The Remiges are divided into **primaries, secondaries,** and **tertials,** according to the location in the wing.

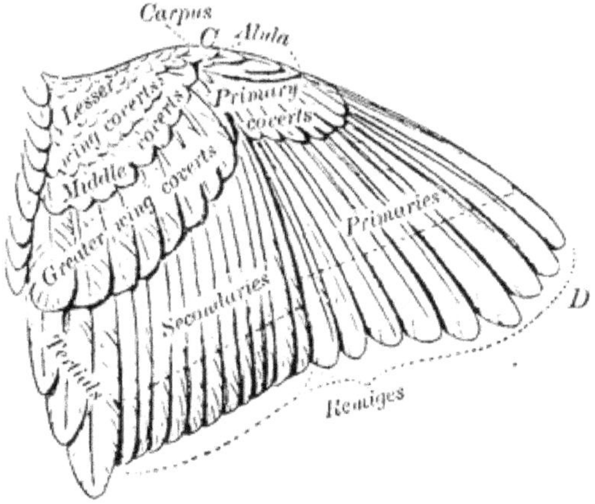

The Primaries are the feathers growing from the outer section of the wing; that is to say, from the outer bend of the wing (**carpus**) to tip, C to D, the number ranging from 9 to 10 (and rarely 11) in various families. At first, it is not always easy to distinguish the last primary from the first secondary; but experience is the best teacher, and the point can always be settled by examining the roots of the feathers.

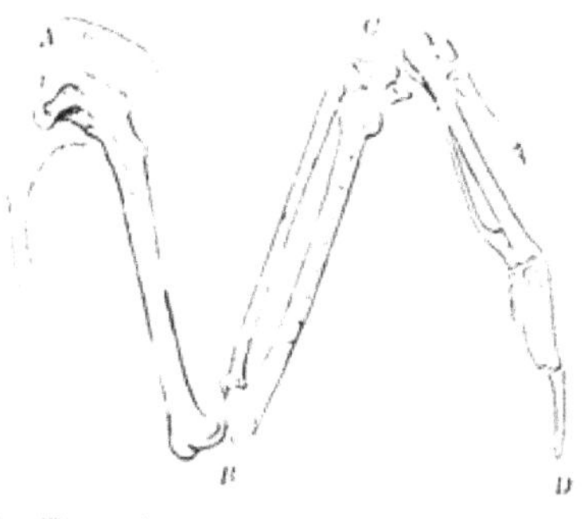

The Secondaries are the **remiges** attached to the *ulna* or forearm, B to C (see illustration); they number from 6 to 40 in the various families; the Humming-bird having the smallest number, and the Albatross more than 40.

The Tertials are the few remaining **remiges** which grow from the *humerus*, A to B.

The Tectrices, or Wing Coverts, are small feathers covering the larger wing feathers; the feathers lining the edge of the under surface of the wing are called **under wing coverts.** The outer wing coverts are divided and described as **greater wing coverts, middle wing coverts,** and **lesser wing** coverts, respectively. (See illustration.)

The Speculum. A term used to indicate a patch or band on the wing (usually of different color from the rest of the feathers), formed by the terminal portion of the secondaries, very noticeable in the Ducks. (See illustration, Topography of a Duck, page 8.)

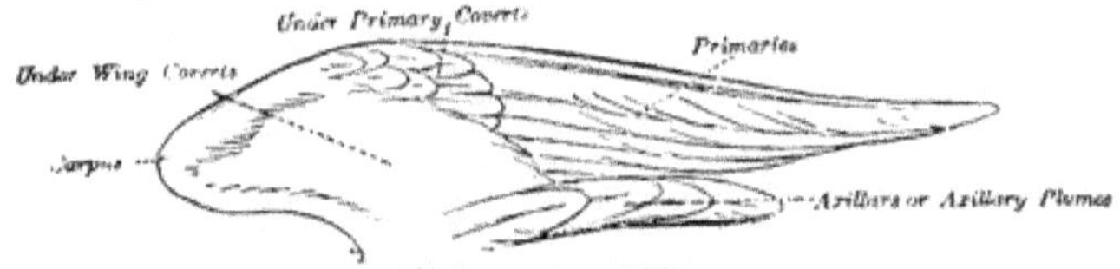

Under Surface of Wing.

The Axillary Plumes, or Axillars, are an important aid in the identification of many species; these are several rather elongated feathers growing from the arm-pit (*axilla*), at the junction of the wing and body.

THE TAIL.

The Rectices, or Tail Feathers, proper, number from 8 to 24, and in some very few cases even more. By far the greater number of birds, however, have 12 rectices.

Truncate or square tail

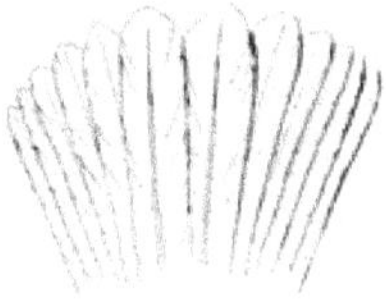

Rounded tail

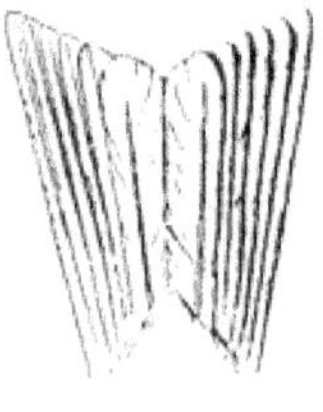

Emarginate tail

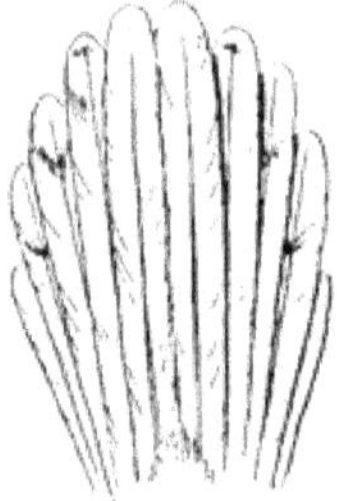

Graduated tail

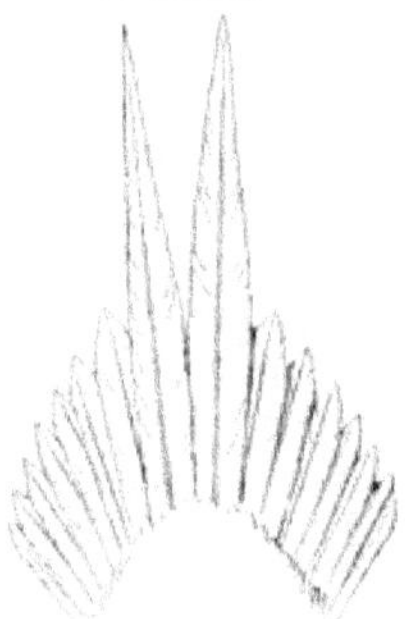

Pointed or cuneate tail

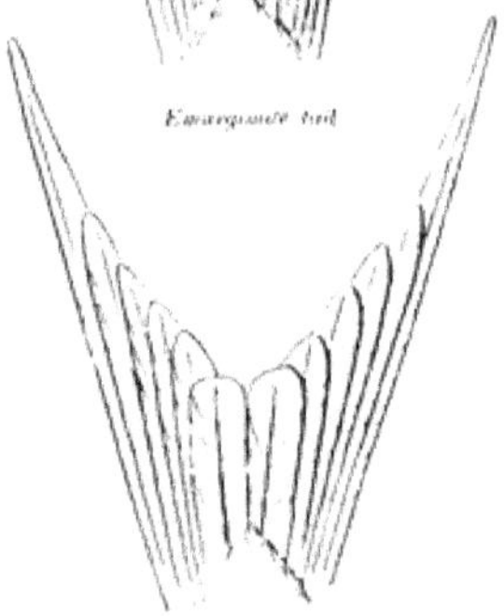

Forked tail

The small feathers overlapping the **rectices** are called **upper tail coverts,** while those below are the **under tail coverts or crissum**. (For illustration of tail coverts and crissum, see Topography of a Bird.)

THE LEG AND FOOT.

A bird's leg may be briefly described as having only the knee downward exposed; the true thigh is concealed, but may be felt under the skin on the side of the body. The true knee is close to the body just under the skin. The first exposed joint which is apparently the knee, but which *bends backward*, is really the heel. (See illustration.) The bird does not walk on the foot (from the heel downward), but merely walks on his toes.

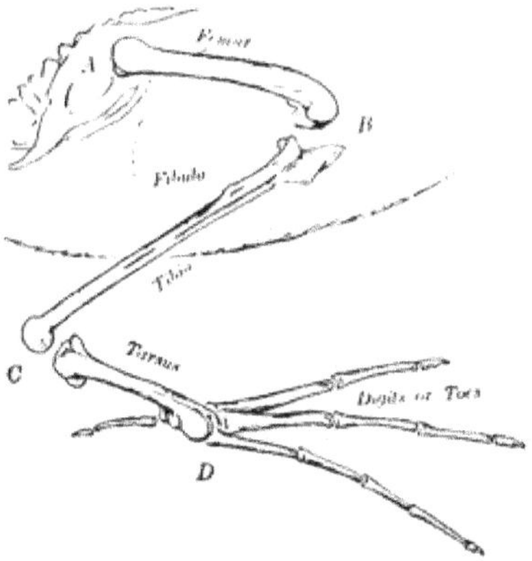

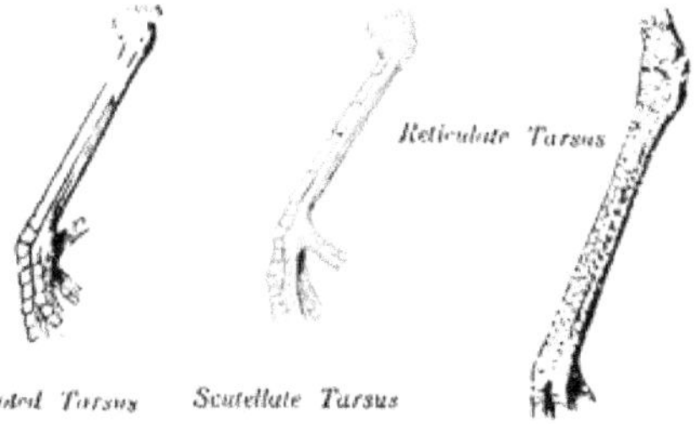

Booted Tarsus *Scutellate Tarsus*

The Tarsus, which is measured from the bend of the leg (end of tibia) to the roots of the toes (C to D in illustration), is an important factor in the identification of many species. The feet are very variable in shape and arrangement of the toes. These variations are of the greatest importance, representing generic and, often, family differences.

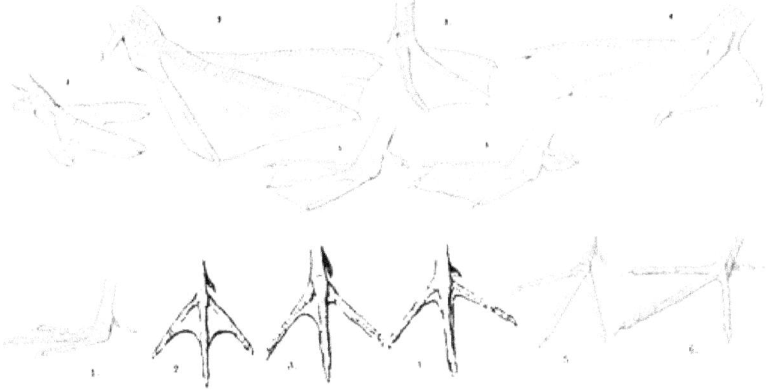

Types of Feet. Water Birds.

The Toes are sometimes three or four (the Ostrich has but two), sometimes two in front and two behind; two in front and one behind; or, in one family (not North America), all four turned forward. Some families have the toes joined or fused together for part of their length, as in the Kingfisher. Some species have webbed feet, others have the toes armed with curved talons, and so on; but the various differences with which we wish to become familiar can better be illustrated than described.

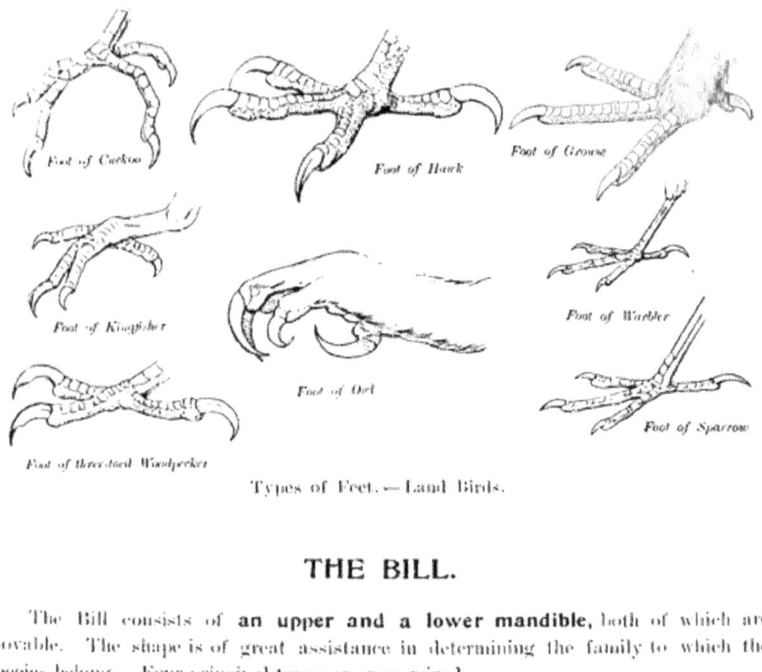

Types of Feet.—Land Birds.

THE BILL.

The Bill consists of **an upper and a lower mandible,** both of which are movable. The shape is of great assistance in determining the family to which the species belong. Four principal types are recognized:—

1. **Epignathous.** Upper mandible longer than lower; the tip bent or hooked over the end of lower mandible. (Examples: Hawks, Gulls, Petrels, Parrots.)
2. **Hypognathous.** Lower mandible, longer than upper mandible. (Examples: Black Skimmers, *Rhynchops.*)
3. **Paragnathous.** Both mandibles of about equal length.
4. **Metagnathous.** Mandibles crossed. (Example: Crossbill. *Loxia.*)

The shape and size of the bill vary greatly, as will be seen by the following illustrations representing fifteen of the principal types among our birds: —

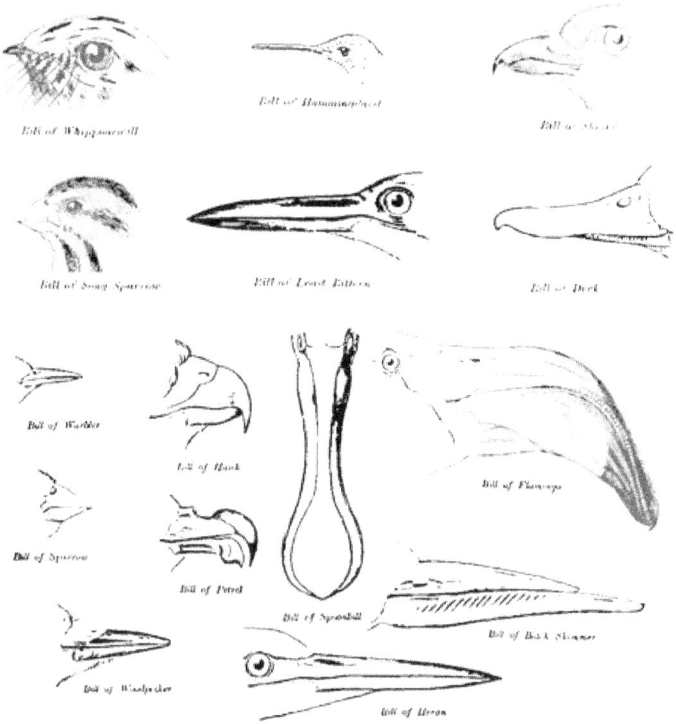

Cere. A membrane (usually hard), which covers the base of the upper mandible. (Hawks, Parrots, Jaegers.)

Nasal Fossa, or Nasal Groove. Groove in which the nostrils open.

Gonys. Lower outline (middle) of the under mandible, A to B.

Culmen. The ridge of the upper mandible.

Unguis. The nail on the end of the upper mandible. Very pronounced in some families of Water Birds, — Ducks, Pelicans, and Petrels.

Mandibles. — Some authors use the word *maxilla* for the upper half of the bill, and *mandible*, for the lower. I prefer, however, to describe the two halves of the bill as upper and lower mandible.

HOW TO MEASURE A BIRD.

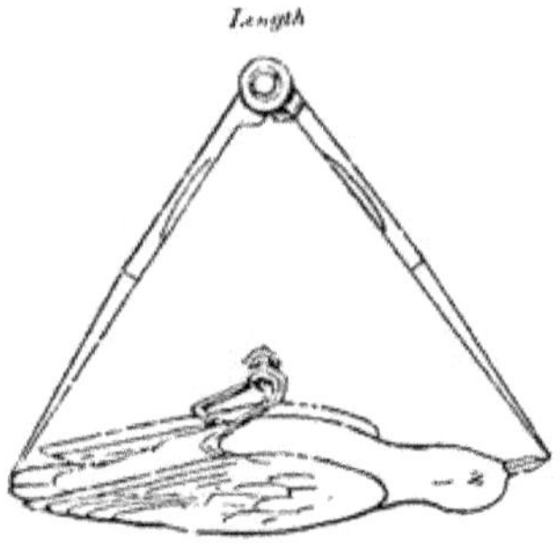

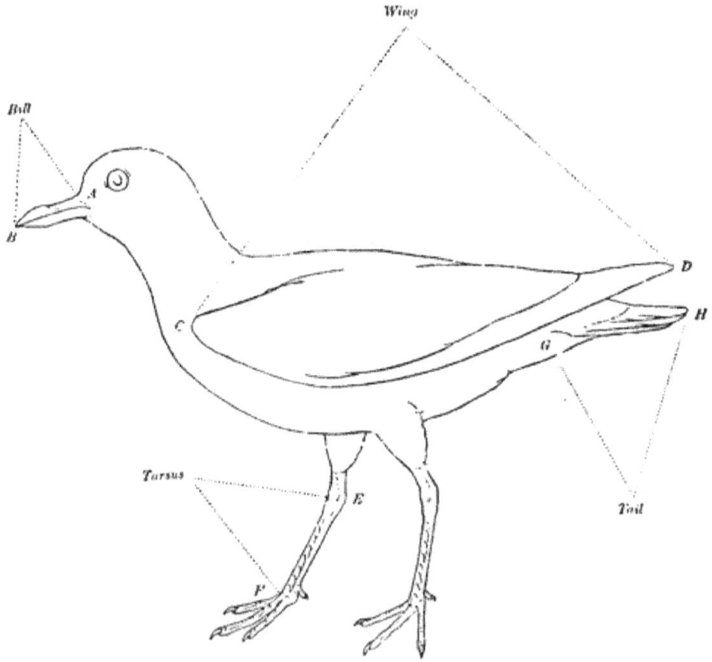

For description see page 9.

HOW TO MEASURE A BIRD.

See Illustration, page 8.

Wing. — Distance from carpal joint C (bend of wing) to the tip of the longest primary D. See cut on page 8.

Length. — Distance in a straight line from the end of the bill to the tip of the longest tail feather. (Occasionally the middle feathers are much elongated, as in the Old Squaw and Pintail Duck, and in other families of birds, such as Phæthon and Stercorarius. In such cases it is well to give the length from bill to longest tail feather, and also to end of outer tail feather.)

Tail. — Distance from the tip of the longest tail feather to its base (the point where it enters the body).

Bill. — The distance in a straight line from where the bill (upper mandible) joins the skin of the forehead (A) to the tip (B). (There are a few exceptions to this rule, in other families, such as birds with frontal plate, etc. Some curved bills are measured along the curve of the culmen, and at times it is advisable to measure from the nostril to the tip of the bill, but in such cases it should always be so stated.)

Tarsus. — Distance in *front* of the leg from what *appears* to be the knee (joint (end of tibia) to the root of the middle toe. All measurements are given in inches and fractions of an inch.

INDEX TO KEY TO FAMILIES.

WATER BIRDS.

Group 1. Toes, four, with lobate web or web on sides of toes. See page 11.

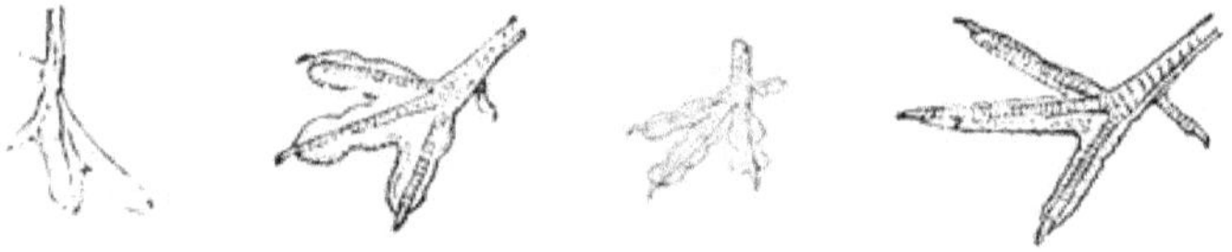

Group 2. Toes, four; front toes, palmate (full webbed); hind toe, not connected with front ones by web. See page 12.

Group 3. Toes, four; toto-palmate (all toes full webbed); hind toe, connected with front ones by web. See page 15.

Group 4. Toes, four, not full webbed; small web between toes at base, or toes entirely without web; hind toe, sometimes very small. See page 17.

Section 1. Hind toe, raised above level of front toes. See page 17.

Section 2. Hind toe, on same level with front toes. See page 19.

Group 5. Toes, three, full webbed; no hind toe. See page 21.

Group 6. Toes, three, not full webbed; a small web between toes at base, or toes entirely without web. See page 23.

KEY TO FAMILIES.

WATER BIRDS.

GROUP 1.

Toes, four, with lobate webs, or webs on sides of toes.

Bill, pointed; feet, placed far back, near tail; underparts, silvery white; tail, very short.

Family PODICIPIDÆ. Grebes.

See page 26.

Forehead, with bare shield; bill, rather short; general color, slaty gray; toes, with lobate webs.

Family RALLIDÆ. Subfamily FULICINÆ. Coots.

See page [illegible]

Hind toe, elevated above the level of the others; bill, slender; nostrils, opening through slits; **sides of toes, webbed.** **Family PHALAROPODIDÆ. Phalaropes.**

See page 104.

GROUP 2.

Toes, four; front toes, palmate (full webbed); hind toe, not connected with front ones by web.

Bill, straight and pointed; tarsus, flattened; hind toe, with flap or lobe; feet, placed far back near the tail; tail, very short.

Family GAVIIDÆ. Loons.

See page 28.

Very large wing, over 19 inches long; upper mandible, curved near tip, forming a hook, the end (unguis) enlarged; nostrils, separate and tubular; hind toe, rudimentary, often apparently wanting.

Family DIOMEDEIDÆ. Albatrosses.

See page 46.

Nostrils, tubular, united in one double-barrelled tube; front toes, palmate (full webbed); hind toe, very small, and in some cases entirely absent; upper mandible, curved near tip; wing, less than 19 inches long.

Family PROCELLARIIDÆ. Shearwaters, Petrels, and Fulmars.

See page 46.

Nostrils, separate not tubular; **bill, with cere** (a horny or skin-like covering on base of upper mandible); hind toe, sometimes very small; end of upper mandible (unguis), swollen and somewhat rounded; back and wings, always dark, sometimes sooty, sometimes barred with brown; tail, never white or gray, usually very dark; **middle tail feathers, longest,**

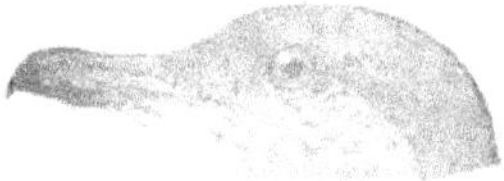
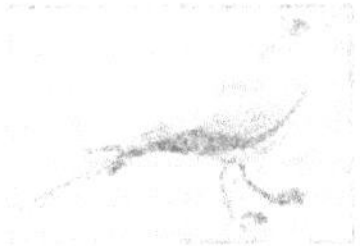

but only very long in adult birds, sometimes only slightly longer than rest of tail feathers in immature birds; in brown plumages, the axillars (feathers extending from armpit) are heavily barred, brown and white; in other plumages, the axillars and under wing coverts are sooty brown or dark slaty brown. **Family STERCORARIIDÆ. Skuas and Jaegers.**

See page 34.

Nostrils, separate not tubular; **bill, without cere**; hind toe, sometimes very small; upper mandible, curved; unguis (end of bill), not swollen; middle tail feathers, about equal in length to the others; tail, rarely dark, although sometimes tipped with black or brown; axillars and under wing coverts, white or gray, sometimes with *narrow* gray lines or faint wavy bars.

Family LARIDÆ. Subfamily LARINÆ. Gulls.

See page 36.

Nostrils, separate not tubular; hind toe, sometimes very small; upper mandible, nearly straight, not hooked or decidedly rounded near tip; outer tail feathers, usually longer than middle feathers. **Family LARIDÆ. Subfamily STERNINÆ. Terns.**

See page 41.

Bill, like blade of a knife, the under mandible the longer; plumage, black above, white below. **Family RHYNCHOPIDÆ. Skimmers.**

See page 45.

Bill, very much bent, with tooth-like serrations on edge. Large, tall birds with long necks and legs, usually red or pink. **Family PHŒNICOPTERIDÆ. Flamingoes.**

See page 85.

Bill, long and narrow, having tooth-like serrations on edges; toes, four, the front ones, full webbed; hind toe, with flap.

Family ANATIDÆ. Subfamily MERGINÆ. Mergansers.

See page 61.

Top of bill of Merganser.

Side of bill.

Bill, not narrow; tarsus, scutellate in front (transverse scales); toes, four, the front ones, full webbed; *hind toes, without decided flap or lobe.*

Family ANATIDÆ. Subfamily ANATINÆ. River and Pond Ducks.

See page 63.

Ducks.

Bill of Duck.

Bill, not narrow; tarsus, scutellate in front (transverse scales); toes, four, the front ones, full webbed; *hind toe, with flap or lobe.*

Family ANATIDÆ. Subfamily FULIGULINÆ. Sea Ducks.

See page 71.

Lores, feathered; tarsus, reticulate (scales rounded); wing, *more than thirteen inches long*; toes, four, the front ones full webbed.

Family ANATIDÆ. Subfamily ANSERINÆ. Geese and Brant.

See page 81.

Geese and Swan.

Lores, partly bare; tarsus, reticulate (scales rounded); size, large; neck, long; wing, over eighteen inches long; plumage, white or gray. **Subfamily CYGNINÆ. Swans.**

See page 84.

GROUP 3.

Toes, four, toto-palmate; (front toes, full webbed;) hind toe, connected with front ones by web.

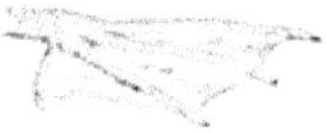

Bill, sharp pointed; chin, feathered; toes, four, all connected by webs.

Family PHAETHONTIDÆ. Tropic Birds.

See page 50.

Bill, stout, but not hooked, chin, bare; neck, thick; toes, four, all connected by webs.

Family SULIDÆ. Gannets.

See page 51.

Bill, sharp-pointed and slender; chin, bare; neck, long and slender; toes, four, all connected by webs; middle tail feathers, corrugated or fluted.

Family ANHINGIDÆ. Darters, Snake Birds.

See page 53.

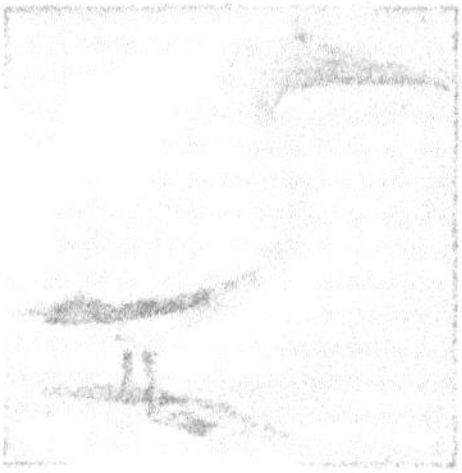

Bill, hooked at tip, over twelve inches long and having a large pouch; lores, bare; toes, four, all connected by webs.

Family PELECANIDÆ. Pelicans

See page 56.

Bill, hooked at tip, and less than twelve inches long; bare skin at base of bill and chin; lores, bare; toes, four, all connected by webs.

Family PHALACROCORACIDÆ. Cormorants.

See page 54.

Bill, hooked at tip; lores, feathered; upper plumage, entirely black; toes, four, all connected by webs; tail, forked; wings, very long.

Family FREGATIDÆ. Man-of-war Birds, Frigate Birds.

See page 58.

GROUP 4.

Toes, four; toes, not full webbed; small webs between toes at base, or toes entirely without webs; hind toe, sometimes small.

Section 1. Hind toe, raised above level of front toes.

Avocet.

Avocet.

Tarsus, over 3.50 inches long; bill, curved upward or straight.

Family RECURVIROSTRIDÆ. Avocets.

See page 106.

Hind toe, higher than front toes; tarsus, less than 3.50 inches long; middle toe and claw together shorter than bill, except a few of the small species, which have middle toe and claw equal to or longer than bill; but all such have the belly and under tail coverts pure white in most plumages. **Family SCOLOPACIDÆ. Snipe, Curlews, Sandpipers, etc.**

See page 107.

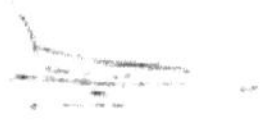

Hind toe, higher than front ones; lower back and rump, **white, with black band.**

Family APHRIZIDÆ. Turnstones.

See page 129.

Hind toe, higher than front ones; hind toe, very small, hardly noticeable; bill, black, rather short and stout; all other species belonging to this family have but three toes.

Family CHARADRIIDÆ (*Charadrius squaterola*). **Black-bellied Plover.**

See page 124.

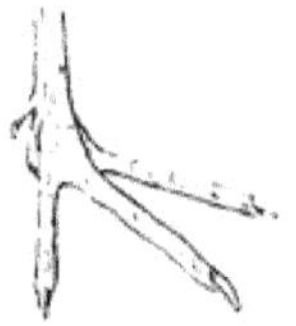

Sora Rail.

Hind toe, above level of front toes; bill, less than 3 inches long; toes, four, no comb-like edge on inner side of middle toe nail; **middle toe and claw together not shorter than bill,** usually decidedly longer; under tail coverts, not white. **Family RALLIDÆ. Rails, etc.**

See page 97.

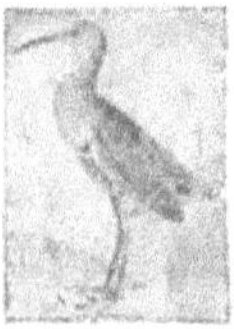

King Rail.

Rails.

Virginia Rail.

Hind toe, above level of front toes; bill, over 3 inches long; **wing, over 16 inches;** tarsus, over 7 inches; toes, four, **no comb-like edge on inner side of middle toe nail;** lores, with hair-like bristles. **Family GRUIDÆ. Cranes.**

See page 95.

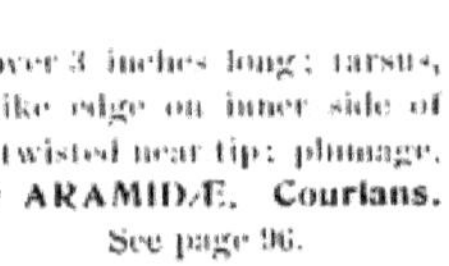

Hind toe, above level of front toes; bill, over 3 inches long; tarsus, under 7; wing, under 16; toes, four, no comb-like edge on inner side of middle toe nail; under mandible, often slightly twisted near tip; plumage, dark brown with white streaks. **Family ARAMIDÆ. Courlans.**

See page 96.

Section 2. Hind toe, on level with front toes.

Toes, long and slender; bill, short and pointed; a bare shield or plate on forehead; wing, about 7 inches long, carpus (bend of wing) to tip; under tail coverts, white.

Family RALLIDÆ. Subfamily GALLINULINÆ.
Purple Gallinule, or Florida Gallinule.

See page 100.

Bill, nearly straight and sharply pointed; inner side of middle toe nail, with distinct comb-like edge; toes, four, all on same level. Bitterns, tail with ten feathers. Herons, tail with twelve feathers.

Family ARDEIDÆ. Herons, Egrets, and Bitterns.

See page 89.

Greater part of plumage, white; bill, rounded and somewhat curved, very thick and strong; *tarsus, always over five inches long*; toes, four, all on same level; *no comb-like edge on inner side of middle toe nail.*

Family CICONIIDÆ. Storks and Wood Ibises.

See page 88.

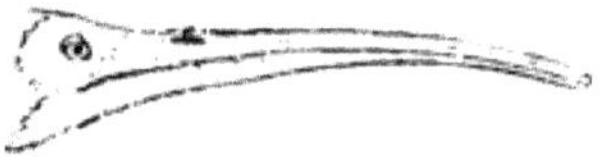

Bill, long, rather slender, and decidedly curved downward; *tarsus, always less than five inches long*; toes, four, all on the same level; no comb-like edge on side of middle toe nail. **Family IBIDIDÆ. Ibises.**

See page 87.

Bill, wide and flat at the end; toes, four, all on same level; wing (carpus to tip), over twelve inches long.

Family PLATALEIDÆ. Spoonbills.

See page 86.

GROUP 5.

Toes, three, full webbed; no hind toe.

Nostrils, separate, not opening into one double-barrelled tube; upper mandible, curved near tip; bill, yellowish, or greenish yellow; **an indication of a hind toe**, in the form of a small knob without nail; tail, entirely white, or white with black band near tip. This is the only North American gull lacking a hind toe.

Family LARIDÆ (*Larus tridactyla*).
Kittiwake Gull.
See page 36.

Bill, reddish (in life showing also yellow and blue), peculiar in form, suggesting that of parrot.
Family ALCIDÆ.
Subfamily FRATERCULINÆ. Puffins.
See page 30.

Bill, black, with narrow white band.
Family ALCIDÆ.
Subfamily ALCINÆ. Auks.
See page 30.

Bill, black or brownish; wing, over 7.25 inches long, carpus (bend of wing) to tip.
Family ALCIDÆ.
Subfamily ALCINÆ.
Murres.
See page 30.

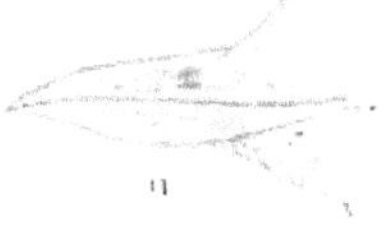

Bill, black, or blackish; wing, less than 7.25, but more than 5.50, measured, carpus to tip.
Family ALCIDÆ. Subfamily PHALARINÆ. Guillemots.
See page 30.

Bill, black, or blackish; wing, less than 5 inches long (carpus to tip).

Family ALCIDÆ. Subfamily ALLINÆ.
Dovekies.
Little Auks.

See page 30.

17

Very large wing, over 19 inches long; upper mandible, curved near tip, forming a hook, the end (unguis) enlarged; nostrils, separate and tubular; hind toe, rudimentary, often apparently wanting.

Family DIOMEDEIDÆ. Albatrosses.

See page 46.

Shearwater. Petrel. Fulmar.

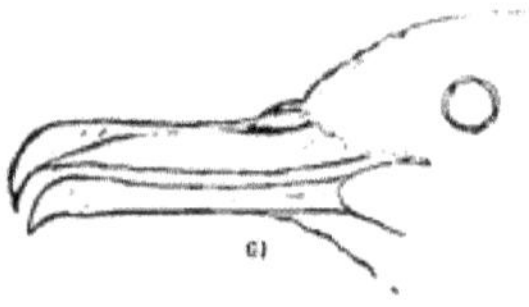

61

Nostrils, tubular, united in one double-barrelled tube; front toes, palmate (full webbed); hind toe, very small, and in some cases entirely absent; upper mandible, curved near tip; wing, less than 19 inches long.

Family PROCELLARIIDÆ.
Shearwaters, Petrels, and Fulmars.

See page 46.

GROUP 6.

Toes, three; a small web between toes, or, entirely without web.

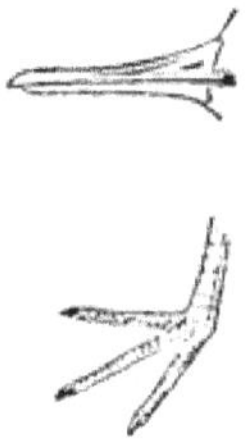

Bill, slender; wing, about 5 inches; bill, about one inch; this is the only representative of this family with three toes, all others have four; tarsus, less than 1.50 inches long.
Family SCOLOPACIDÆ. (*Calidris arenaria.*) **Sanderling. Sandpiper.**
See page 107.

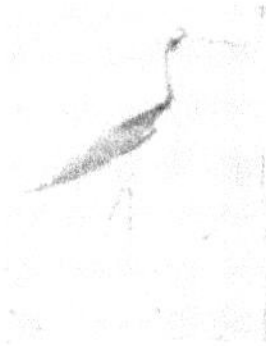

Toes, three, partly webbed; tarsus, over 3 inches long; general plumage, black and white; legs, pink red in life. **Family RECURVIROSTRIDÆ. Stilts.**
See page 106.

Plovers.

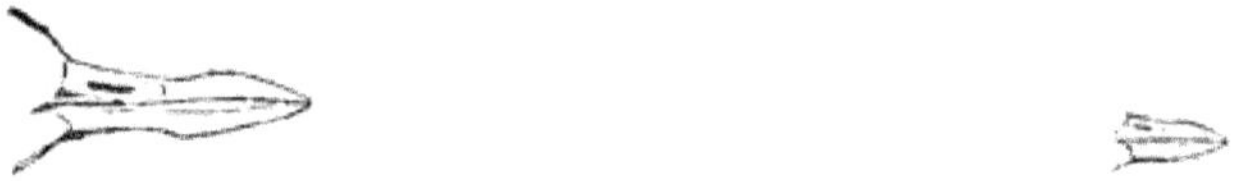

Bill, short; some species have bill less than three quarters inch long; none have bill over two inches long.

Family CHARADRIIDÆ. Plovers.

See page 124.

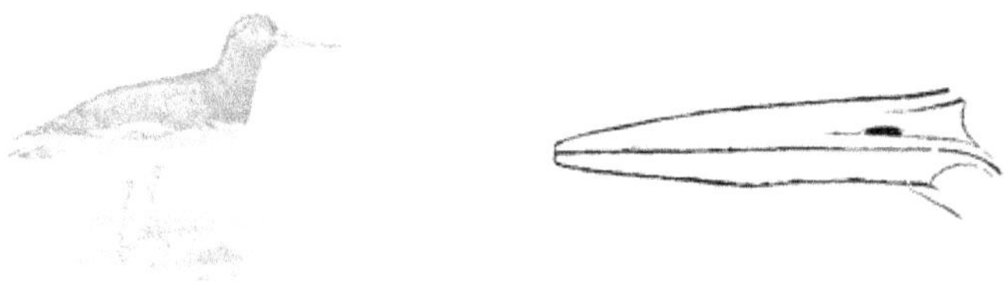

Bill, over 2.50 inches long; head and neck, black or blackish; bill, red.

Family HÆMATOPODIDÆ. Oyster-catchers.

See page 130.

WATER BIRDS.

KEY TO THE SPECIES.

FAMILY PODICIPIDÆ.

Grebes.

Toes, four; three in front, one behind, with lobate webs.

* Group 1. Wing, less than 6.50 inches long.

Depth of bill at base of culmen, over .40. *In summer:* Bill, with black band; throat, black.

In winter: Bill, without black band; throat, white.

Podilymbus podiceps. **Pied-billed Grebe. Hell Diver.**
See No. 4.

Depth of bill at base of culmen, less than .40. *In summer:* Two small tufts of brownish gray or buff-colored feathers behind the eye; crown, nape, and throat, black.

In winter: Plain colored, without black or buff on head; upper plumage, grayish black; underparts, silvery white, often tinged with ash gray on throat and sides.

Colymbus auritus. **Horned Grebe.**
See No. 2.

* Group 2. Wing, over 6.50 inches long.

In summer: Crown, black; upper throat, gray; lower throat and breast, chestnut rufous; rest of underparts, silvery grayish white.

In winter: No rufous brown on throat or breast. *Colymbus holboellii.* **Holboell's Grebe.**
See No. 3.

Grebes.

Horned Grebe.

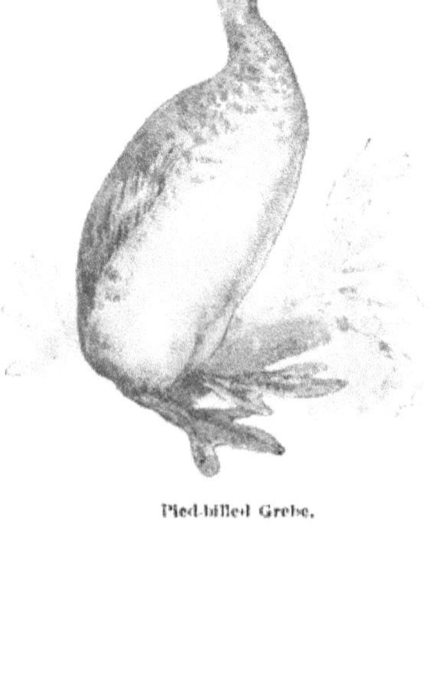

Pied-billed Grebe.

FAMILY GAVIIDÆ.

Loons.

Toes, four; front toes, palmate (full webbed); bill, pointed; tarsus, flattened; hind toe, with flap or small lobe.

* Group 1. Wing, 9.50 to 11 inches long.

Back, with white spots; throat, white or whitish (immature or winter).
Gavia lumme, **Red-throated Loon.**
See No. 7.

Back, without distinct white spots; the feathers, **edged** with ashy; throat, white or dusky (immature or winter).
Gavia arctica, **Black-throated Loon.**
See No. 6.

Head, ashy gray; throat, black (adult). *Gavia arctica*, **Black-throated Loon.**
See No. 6.

Throat, gray; front of neck, chestnut brown (adult).
Gavia lumme, **Red-throated Loon.**
See No. 7.

* Group 2. Wing, 11 to 15 inches long.

Section 1. Depth of bill at base, more than .85.

Head, black (adult). *Gavia imber*, **Loon.**
See No. 5.

Top of head, grayish; throat, white (immature or winter plumage).
Gavia imber, **Loon.**
See No. 5.

Section 2. Depth of bill at base, less than .85.

Head, ash gray; throat, black (adult). *Gavia arctica*, **Black-throated Loon.**
See No. 6.

Throat, gray; front of neck, chestnut brown (adult).
Gavia lumme, **Red-throated Loon.**
See No. 7.

* For directions for measurement, see page [illegible].

Summer. Loons. Winter.

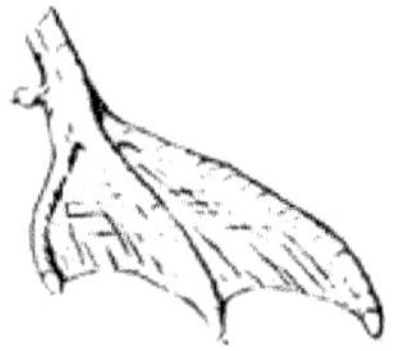

Loon.

Black-throated Loon.

Back, **with white spots ;** throat, white or whitish (immature or winter).
Gavia lumme. **Red-throated Loon.**
See No. 7.

Back, **without distinct white** spots; the feathers, **edged** with ashy; throat, white or dusky (immature or winter). *Gavia arctica.* **Black-throated Loon.**
See No. 6.

* Group 3. Wing, over 15 inches long.

Feet, webbed; head, black, in adult; head, gray; throat, whitish, in immature.
Urinator imber. **Loon.**
See No. 5.

FAMILY ALCIDÆ.

Auks, Puffins, and Murres.

Toes, three ; no hind toe (hallux) ; toes, palmate (full webbed) ; nostrils, separate, and not tubular.

* Group 1. Wing, less than 5 inches long.

Winter plumage : Upper parts, black; underparts, white. *In summer :* The breast, throat, and sides of the head and neck, sooty brown.

Alle alle. **Dovekie. Little Auk.**
See No. 17.

* Group 2. Wing, 5.50 to 6.50 inches long.

Bill, parrot like; underparts, white; back, black; usually a black collar. *Fratercula arctica.* **Puffin. Sea Parrot.**
See No. 9.

Adult in summer : **General plumage, black,** a white patch on the shoulder; **basal half of the greater wing coverts** (feathers forming the white patch on the wing), **black.** Winter birds have the underparts white, and the upper parts, black and white.
Cepphus grylle. **Black Guillemot. Sea Pigeon.**
See No. 11.

Similar to Black Guillemot, but has the **wing coverts** (feathers forming the white wing patch) **entirely white** and **not with basal half black.**

Immature and winter birds (except wing coverts), resembling the Black Guillemot.
Cepphus mandtii. **Mandt's Guillemot.**
See No. 12.

* For directions for measurement, see page [illegible].

Murres.

Puffin. Guillemots.

* Group 3. Wing, 6.50 to 7.50 inches long.

Bill, parrot like; underparts, white; back, black, usually a black collar. *Fratercula arctica, and races.* **Puffin. Sea Parrot.**
See No. 9.

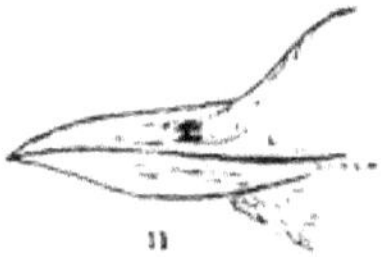

Adult in summer: **General plumage, black;** a white patch on the shoulder; **basal half of the greater wing coverts** (feathers forming the white patch on the shoulder), **black.** Winter birds have the underparts white, and the upper plumage, mixed black, gray, and white. *Cepphus grylle.* **Black Guillemot.**
See No. 11.

Similar to the Black Guillemot, but has the **wing coverts entirely white** and not with basal half black. *Cepphus mandtii.* **Mandt's Guillemot.**
See No. 12.

Summer plumage: Head, back, wings, and tail, sooty brown; underparts and tips of secondaries, white. *Winter plumage:* Underparts, white, more or less marked with sooty brown, or blackish about the throat, belly, and flanks. *Uria lomvia.* **Brunnich's Murre.**
See No. 14.

Summer plumage: Head and neck, back, wings, and tail, black; tips of secondaries and rest of underparts, white. *Winter plumage:* Similar, but has the throat white. *Alca torda.* **Razor-billed Auk.**
See No. 15.

* Group 4. Wing, 7.50 to 8.50 inches long.

Summer plumage: Head and neck, back, wings, and tail, black. Tips of secondaries and rest of underparts, white. *Alca torda.* **Razor-billed Auk.**
See No. 15.

Summer plumage: Head, back, wings, and tail, dark sooty brown; underparts and tips of secondaries, white. *Winter plumage:* Underparts, white, more or less marked with sooty brown or blackish about the throat, belly, and flanks; bill, usually over 1.60.
Uria troile. **Murre.**
See No. 13.

Similar to *Uria troile*, but has the head darker than the throat; bill, usually under 1.60.
Uria lomvia. **Brunnich's Murre.**
See No. 14.

* For directions for measurement see page [illegible]

* Group 5. Wing, over 8.50 inches long.

Auk.

Head and neck, **black,** easily distinguished by the shape of the bill; a white line on bill.
Alca torda. **Razor-billed Auk.**
See No. 15.

Murre.

Auk.

Head and neck, sooty **brown;** bill pointed, no white line on bill. *Uria troile.* **Murre.**
See No. 13.

* For directions for measurement, see page 9.

FAMILY STERCORARIIDÆ.

Skuas and Jægers.

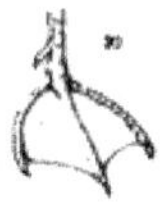

Nostrils, separate, and not tubular; bill, with cere; front toes, palmate (full webbed); hind toe, small, but always present; end of upper mandible (unguis), swollen, and somewhat rounded; back and wings, always dark, sometimes sooty, sometimes barred with brown; tail, never white or gray, usually very dark; middle tail feathers, longest. The cere (hard skin-like covering on base of upper mandible) will distinguish these birds from Gulls.

* Group 1. Wing, 11 to 15 inches long.

Underparts, not entirely pure white; bill, over 1.35; tarsus, over 1.80; middle tail feathers, not pointed; plumage, sometimes dark brown; sometimes mixed gray, brown, and white. *Stercorarius pomarinus.* Pomarine Jæger.
See No. 19.

Bill, under 1.35; tarsus, under 1.80; middle tail feathers, pointed; base of unguis to frontal feathers less than length of unguis; shafts of primaries, yellow white. *Stercorarius longicaudus.* Long-tailed Jæger.
See No. 21.

Bill, under 1.35; tarsus, under 1.80; middle tail feathers, pointed; base of unguis to frontal feathers, greater than length of unguis; shafts of primaries, yellowish white. *Stercorarius parasiticus.* Parasitic Jæger.
See No. 20.

* Group 2. Wing, over 15 inches long.

Bill, with cere; general color, dark brown, more or less streaked with light brown; a patch of white at base of primaries. *Megalestris skua.* Skua.
See No. 18.

Skua.

Jaeger.

Jaeger.

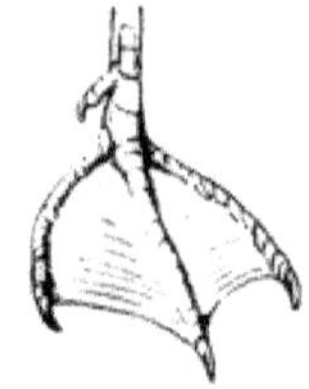

FAMILY LARIDÆ.

GULLS AND TERNS.

SUBFAMILY LARINÆ. GULLS.

TAIL, USUALLY NEARLY SQUARE.

GROUP 1. Wing, 9.50 to 11 inches long. See page 36.
GROUP 2. Wing, 11 to 13 inches long. See page 36.
GROUP 3. Wing, 13 to 15 inches long. See page 38.
GROUP 4. Wing, 15 to 17 inches long. See page 38.
  Section 1. Length, over 21; bill, over 2; underparts, white. See page 38.
  Section 2. Length, over 21; bill, over 2; underparts, not white. See page 39.
  Section 3. Length, over 21; bill, under 2; underparts, white. See page 39.
  Section 4. Length, over 21; bill, under 2; underparts, not white. See page 40.
  Section 5. Length, under 21. See page 40.
GROUP 5. Wing, over 17 inches long. See page 40.

SUBFAMILY STERNINÆ. TERNS.

TAIL, USUALLY FORKED.

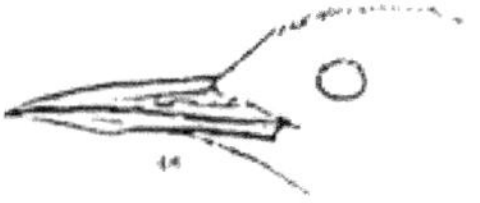

GROUP 1. Wing, 5.50 to 7.50 inches long. See page 41.
GROUP 2. Wing, 7.50 to 8.50 inches long. See page 41.
GROUP 3. Wing, 8.50 to 9.50 inches long. See page 41.
GROUP 4. Wing, 9.50 to 11 inches long. See page 41.
GROUP 5. Wing, 11 to 13 inches long. See page 43.
GROUP 6. Wing, 13 to 17 inches long. See page 43.
GROUP 7. Wing, over 17 inches long. See page 43.

FAMILY LARIDÆ.

Gulls and Terns.

SUB-FAMILY LARINÆ.

Gulls.

Upper mandible, curved; unguis (end of bill), not swollen; middle tail feathers, about equal in length to the others; tail, rarely dark, although sometimes tipped with black or brown; hind toe, small, but always present except in one genus; bill, without cere.

* Group 1. Wing, 9.50 to 11 inches long.

Underparts, white; inner web of first primary, **white,** with black spot near the end; the **tip, white; shaft of feather, white;** adult birds have the bill dark red; in immature birds it is brownish. *Larus franklinii.* **Franklin's Gull.**
See No. 34.

Underparts, white; inner web of first primary, **white;** the **tip, black;** bill, black; shaft of **feather, white.** *Larus philadelphia.* **Bonaparte's Gull.**
See No. 35.

Underparts, white; inner web of first primary, about **half white; shaft of feather, dark brown on upper surface.** In the adult the bill is black, **tipped with yellow.** *Xema sabinii.* **Sabine's Gull.**
See No. 38.

* Group 2. Wing, 11 to 13 inches long.

Shafts of primaries, black or dark brown; underparts, not pure white (immature). *Larus atricilla.* **Laughing Gull.**
See No. 33.

Underparts, pure white; inner web of first primary, **white,** with black spot near the end, **the tip, white;** shaft of feathers, white. Adult birds have bill dark red; in immature birds it is brownish. *Larus franklinii.* **Franklin's Gull.**
See No. 34.

* For directions for measurement, see page 9.

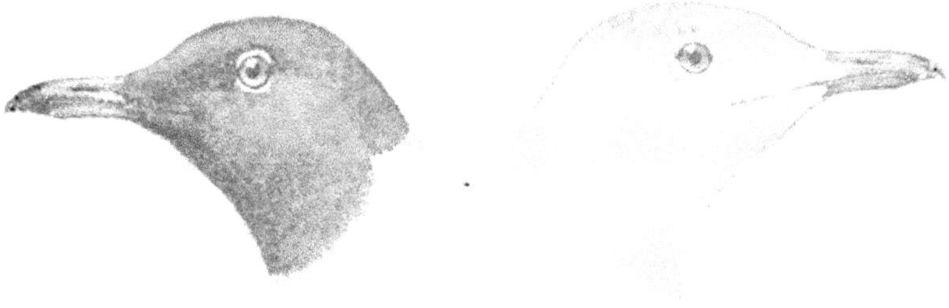

Summer. Laughing Gull. Winter.

Ivory Gull.

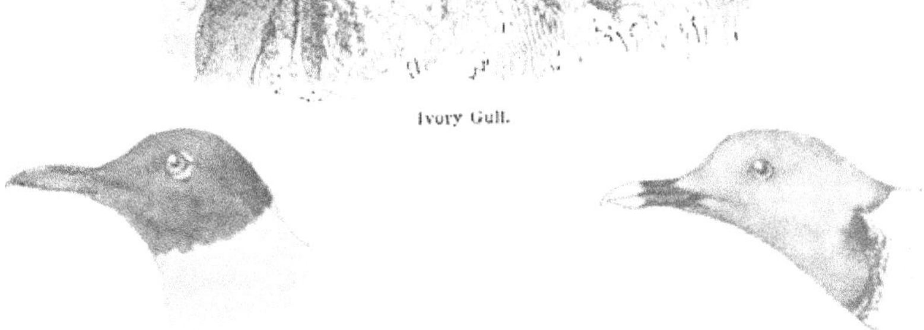

Underparts, white; first primary, entirely black or dark brown (adult).
Larus atricilla. **Laughing Gull.**
See No. 33.

General plumage, white: back, **white;** first primary entirely white, or with a blackish spot at tip; **hind toe, with nail;** bill, yellow. *Pagophila alba.* **Ivory Gull.**
See No. 22.

Back, pearl gray; inner web of primary, white, broadly **tipped** with black; a small rudimentary hind toe **without nail.** *Rissa tridactyla.* **Kittiwake Gull.**
See No. 23.

* Group 3. Wing, 13 to 15 inches long.

Underparts, white; first primary entirely black, or dark brown.
Larus atricilla. **Laughing Gull.**
See No. 33.

General plumage, white; first primary, entirely white, or with a dusky spot near tip; hind toe, with nail; bill, less than 1.50; **wing, less than 14;** bill, yellow.
Pagophila alba. **Ivory Gull.**
See No. 22.

Underparts, white; inner web of primary, **white, broadly tipped** with black; a small rudimentary hind toe, **without nail.** *Larus tridactyla.* **Kittiwake Gull.**
See No. 23.

Underparts, white; terminal portion of first primary, **black,** with white spot near tip.
Larus delawarensis. **Ring-billed Gull.**
See No. 31.

Primaries, pale pearl gray, becoming white at tip; bill, over 1.50; **wing, over 14.**
Larus leucopterus. **Iceland Gull.**
See No. 25.

* Group 4. Wing, 15 to 17 inches long.

Section 1. Length, over 21; bill, over 2; underparts, white.

No black on primaries (adult). *Larus glaucus.* **Glaucous Gull.**
See No. 24.

Back, slaty black; primaries, black with more or less white (adult).
Larus marinus. **Great Black-backed Gull.**
See No. 27.

Back, pale bluish, gray, or pearl gray; primaries, dull black or brownish black, more or less marked with white.
Larus argentatus smithsonianus. **American Herring Gull.**
See No. 30.

* For directions for measurement, see page 9.

Section 2. Length, over 21; bill, over 2; underparts, not white.

Outer webs of primaries, ash color (immature). *Larus glaucus.* **Glaucous Gull.**
See No. 24.

Outer webs of primaries, dark brown; wing, over 17.50; depth of bill at angle, over .90 (immature). *Larus marinus.* **Great Black-backed Gull.**
See No. 27.

Outer webs of primaries, dark brown; depth of bill at angle, under .90; wing, under 17.50 (immature). *Larus argentatus smithsonianus.* **American Herring Gull.**
See No. 30.

Section 3. Length, over 21; bill, under 2; underparts, white.

Back, grayish blue, more or less black on primaries. *Larus argentatus smithsonianus.* **American Herring Gull.**
See No. 30.

Back, pale pearl color; primaries, whitish or pearl color, shading to white at tips. *Larus leucopterus.* **Iceland Gull.**
See No. 25.

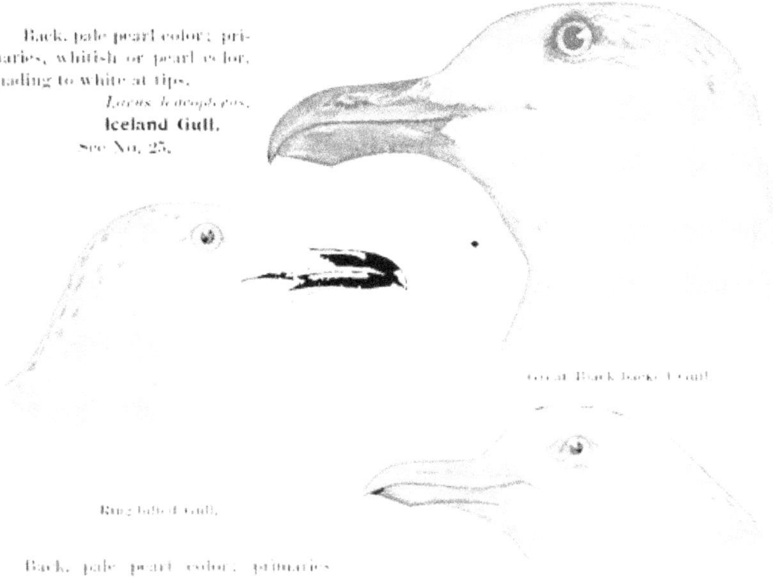

Great Black-backed Gull.

Ring-billed Gull.

Back, pale pearl color; primaries marked with slaty gray. *Larus kumlieni.* **Kumlien's Gull.**
See No. 26.

Section 4. Length, over 21; bill, under 2; underparts, not white.

Outer webs of primaries, dark brown; bill, over 1.90 (immature).
Larus argentatus smithsonianus. **American Herring Gull.**
See No. 30.

Outer webs of primaries, brownish; bill, under 1.90 (immature).
Larus kumlieni. **Kumlien's Gull.**
See No. 26.

Outer webs of primaries, ash color (immature). *Larus leucopterus.* **Iceland Gull.**
See No. 25.

Section 5. Length, under 21.

A band of black on the bill (adult). Bill, dull yellow, tipped with black (immature).
Larus delawarensis. **Ring-billed Gull.**
See No. 31.

Terns. Gulls.

* Group 5. Wing, over 17 inches long.

Primaries, marked with more or less black. *Larus marinus.* **Great Black-backed Gull.**
See No. 27.

No black on primaries. *Larus glaucus.* **Glaucous Gull.**
See No. 24.

* For directions for measurement, see page 9.

SUB-FAMILY STERNINÆ.

Terns.

Upper mandible, nearly straight, not hooked or decidedly rounded near tip; outer tail feathers, usually longer than middle feathers; toes, four; front toes, webbed; hind toe, small, but well developed.

* Group 1. Wing, 5.50 to 7.50 inches long.

48

Forehead, white; bill, black; underparts, white; back, pearl gray; crown, black (adult in summer); back and crown, mottled (immature). *Sterna antillarum.* **Least Tern.**
See No. 48.

* Group 2. Wing, 7.50 to 8.50 inches long.

Head and underparts, black (adult). A patch of black behind the eye; back of head, dusky; rest of head, white (immature). *Hydrochelidon nigra surinamensis.* **Black Tern.**
See No. 51.

* Group 3. Wing, 8.50 to 9.50 inches long.

Outer tail feathers, pure white. *Sterna dougalli.* **Roseate Tern.**
See No. 47.

Inner web of outer tail feather, gray. *Sterna forsteri.* **Forster's Tern.**
See No. 44.

* Group 4. Wing, 9.50 to 11 inches long.

Outer web of outer **tail feather, darker** than **inner web;** back, pearl gray; breast, washed with pearl gray; bill in adults, **red,** tipped with black; tarsus, usually over .70.
Sterna hirundo. **Common Tern. Wilson's Tern.**
See No. 45.

Inner web of outer tail feather, darker than outer web; back, pearl gray; breast, white; bill in adults, **black,** slightly yellowish at tip. *Sterna forsteri.* **Forster's Tern.**
See No. 44.

Outer tail feather, entirely pure white; back, pearl gray; breast, white, often tinged with faint rose color; bill, in adults, black with basal portion, red.
Sterna dougalli. **Roseate Tern.**
See No. 47.

* For directions for measurement, see page

Forster's Tern. Winter.

Roseate Tern.

Common Tern.

Least Tern.

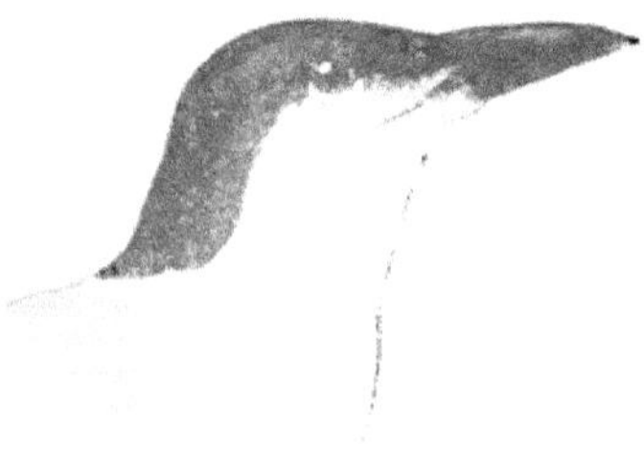

Gull-billed Tern.

Adult Terns (except the Noddy) have the top of the head entirely black, part of the year, varying with age and season.

Outer web of outer tail feather, darker than inner web; back, pearl gray; breast and underparts, pearl gray; bill, in adults, entirely red; tarsus, usually less than .70.
Sterna paradisæa. **Arctic Tern.**
See No. 46.

Back, black; a white stripe from forehead, **extending over the eye;** breast, white; bill, in adults, black. *Sterna fuliginosa.* **Sooty Tern.**
See No. 49.

Back, sooty or grayish brown; a white stripe from forehead, **not reaching above the eye;** bill, black. *Sterna anæthetus.* **Bridled Tern.**
See No. 50.

Back, sooty brown; underparts, sooty brown. *Anous stolidus.* **Noddy Tern.**
See No. 53.

* Group 5. Wing, 11 to 13 inches long.

Bill, black, not tipped with yellow; feet, blackish; **back, pearl gray;** bill, comparatively short and stout. *Gelochelidon nilotica.* **Gull-billed Tern.**
See No. 39.

Bill, black, not tipped with yellow; feet, black; **back, black or sooty.**
Sterna fuliginosa. **Sooty Tern.**
See No. 49.

Bill, red, tipped with black; feet, orange red; back, pearl gray or gray and buff.
Sterna hirundo. **Common Tern. Wilson's Tern.**
See No. 43.

Bill, black, tipped with pale yellow; feet, dark; bill, comparatively long and slender.
Sterna sandvicensis acuflavida. **Cabot's Tern.**
See No. 39.

* Group 6. Wing, 13 to 17 inches long.

Bill, orange or yellowish; tarsus, less than 1.50; inner web of outer primary, usually with more or less white. *Sterna maxima.* **Royal Tern.**
See No. 41.

Bill, red or reddish; tarsus, over 1.50; inner web of outer primary, **usually** without white. *Sterna caspia.* **Caspian Tern.**
See No. 40.

Bill, black, tipped with yellow. *Sterna sandvicensis acuflavida.* **Cabot's Tern.**
See No. 39.

* Group 7. Wing, over 17 inches long.

Bill, red or reddish. *Sterna caspia.* **Caspian Tern.**
See No. 40.

* For directions for measurement, see page 9.

Caspian Tern.

Royal Tern.

Top of head is at times entirely black

Bridled Tern.

Sooty Tern.

Noddy Tern.

FAMILY RHYNCHOPIDÆ.

Skimmers.

Bill, like blade of a knife, the under mandible the longer; plumage, black above, white below.

* Group 1. Wing, 13 to 15 inches long.

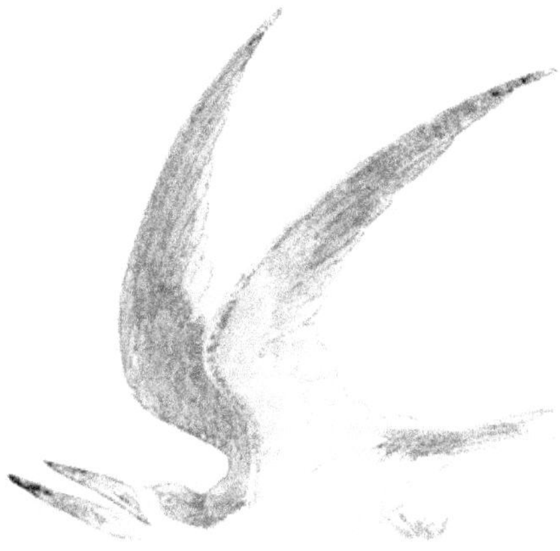

FAMILY DIOMEDEIDÆ.

Albatrosses.

Very large wing, over 19 inches long; upper mandible, curved near tip, forming a hook, the end (unguis) enlarged; nostrils, separate and tubular; hind toe, rudimentary, often apparently wanting.

* Group 1. Wing, 17 to 21 inches long

Bill, dark; the top (culmen), yellow. A very large sea bird.

Thalassogeron culminatus. **Yellow-nosed Albatross.**
See No. 56.

FAMILY PROCELLARIIDÆ.

Shearwaters, Petrels, and Fulmars.

Nostrils, tubular, united in one double-barrelled tube; front toes, palmate (full webbed); hind toe, very small, and in some cases entirely absent; upper mandible, curved near tip; wing, less than 19 inches long.

* Group 1. Wing, less than 5.50 inches long.

General plumage, sooty black, the underparts with faint brownish tinge; upper tail coverts, white, some of them tipped with black. *Procellaria pelagica.* **Stormy Petrel.**
See No. 67.

67

* Group 2. Wing, 5.50 to 6.50 inches long.

Tail, nearly square; upper tail coverts, white, not tipped with black; underparts, dull black; **tarsus, over 1.05.** *Oceanites oceanicus.* **Wilson's Petrel.**
See No. 69.

Tail, decidedly forked; upper tail coverts, white, not tipped with black; underparts, brownish black or sooty brown; **tarsus, less than 1.05.**
Oceanodroma leucorhoa. **Leach's Petrel.**
See No. 68.

Underparts, white. *Pelagodroma marina.* **White-faced Petrel.**
See No. 71.

For directions for measurement see page 3.

Leach's Petrel.

Stormy Petrel.

Wilson's Petrel.

* Group 3. Wing, 6.50 to 9.50 inches long.

Upper parts, dark sooty brown or grayish black; underparts, white; sides of breast, tinged with gray; **middle toe and claw, less than 2 inches long.**

Puffinus auduboni. **Audubon's Shearwater.**
See No. 62.

The Manx Shearwater, *Puffinus puffinus*, a European species, somewhat resembles Audubon's Shearwater, but it is larger, the back darker, the wing rarely, if ever, measuring less than 8.40 inches, and the middle toe and claw 2 inches or more. It is of accidental occurrence on our coast.

* Group 4. Wing, 11 to 13 inches long.

Upper and under tail coverts, gray or brownish gray; breast, white.
Puffinus major. **Greater Shearwater.**
See No. 60.

Under tail coverts, gray; upper tail coverts, sooty; breast, gray.
Puffinus fuliginosus. **Sooty Shearwater.**
See No. 63.

Upper and under tail coverts, white. *Æstrelata hasitata.* **Black-capped Petrel.**
See No. 64.

Upper tail coverts, pearl gray; under tail coverts, white, or entire plumage, dark slate color (dark phase). *Fulmarus glacialis, and races.* **Fulmar Petrel.**
See Nos. 57 and 58.

* Group 5. Wing, 13 to 15 inches long.

Upper surface of tail feathers, dark slaty brown; under tail coverts, ash gray or **brownish gray.** *Puffinus major.* **Greater Shearwater.**
See No. 60.

Upper surface of tail feathers, dark slaty brown; under tail coverts, **white,** sometimes slightly mottled with grayish. *Puffinus borealis.* **Cory's Shearwater.**
See No. 59.

Upper surface of tail feathers, pale pearl gray; upper tail coverts, pearl gray; under tail coverts, **white;** or entire plumage, dark slate color (dark phase).
Fulmarus glacialis, and races. **Fulmar Petrel.**
See Nos. 57 and 58.

* Group 6. Wing, 15 to 17 inches long.

Upper surface of tail feathers, dark slaty brown; **under tail coverts, ash gray,** or brownish gray. *Puffinus major.* **Greater Shearwater.**
See No. 60.

Upper surface of tail feathers, dark slaty brown; under tail coverts, **white;** sometimes slightly mottled with grayish. *Puffinus borealis.* **Cory's Shearwater.**
See No. 59.

* For directions for measurement, see page 9.

Fulmar.

Cory's Shearwater.

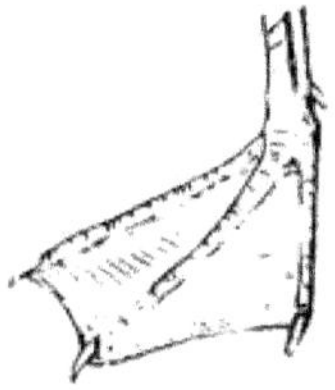

Puffinus major.

Audubon's Shearwater

Black-capped Petrel.

ORDER STEGANOPODES.

Gannets, Pelicans, Cormorants, etc.

Toes, four, all connected by webs.

FAMILY PHAETHONTIDÆ.

Tropic Birds.

Bill, sharp pointed; chin, feathered; toes, four, all connected by webs.

* Group 1. Wing, 9.50 to 13 inches long.

Bill, yellowish; general plumage, white; outer webs of primaries and lesser wing coverts, black; middle tail feathers, very long; bill, yellow or pale orange; shafts of tail feathers, black. *Phaëthon americanus*, **Yellow-billed Tropic Bird.**
See No. 72.

Bill, red; back, finely barred with black. Rare straggler, recorded from Newfoundland banks. *Phaëthon æthereus*. **Red-billed Tropic Bird.**
See No. 73.

* For directions for measurement, see page 9.

FAMILY SULIDÆ.

Gannets.

Bill, stout, but not hooked; chin, bare; neck, thick; toes, four, all connected by webs.

* Group 1. Wing, 13 to 15 inches long.

Head, sooty brown; belly, white; feet, yellowish or greenish, but never red (adult).

Sula sula. **Booby Gannet.**
See No. 75.

Head, sooty brown; belly, brownish; feet, not red (immature).

Sula sula. **Booby Gannet.**
See No. 75.

Head, white, tinged with buff; belly, white; **feet, red** (adult).

Sula piscator. **Red-footed Booby.**
See No. 76.

* Group 2. Wing, 15 to 17 inches long.

Section 1. Side of chin, feathered; a narrow strip of bare skin down the middle of the throat.

Adult, white head, tinged with buff; primaries, dark; immature birds are mottled, grayish brown and white. *Sula bassana.* **Gannet.**
See No. 77.

Section 2. Whole of chin and upper part of throat, bare skin without feathers.

General plumage, sooty brown (rarely whitish); underparts, **white; feet, greenish yellow, or pale yellow.** *Sula sula.* **Booby Gannet.**
See No. 75.

General plumage, white; **feet, red;** gular sack, blackish; **bare skin in front of eye and angle of jaw, pink red;** tail, white; immature birds are grayish brown.

Sula piscator. **Red-footed Booby.**
See No. 76.

General plumage, white; **feet, leaden blue;** gular sack, blackish; bare skin in front of eye, bluish; tail, sooty brown; the middle feathers tinged with hoary; young birds are sooty gray. *Sula cyanops.* **Blue-faced Booby.**
See No. 74.

* Group 3. Wing, 17 to 21 inches long.

Sides of chin, feathered; a narrow strip of bare skin down middle of throat; adults, white, head tinged with buff; immature birds, mottled grayish brown and white. *Sula bassana.* **Gannet.**
See No. 77.

Whole chin and upper throat, bare; feet, leaden blue; wing, always less than 18. Accidental straggler, on Florida coast, not recorded elsewhere in Eastern North America.

Sula cyanops. **Blue-faced Booby Gannet.**
See No. 74.

* For directions for measurement, see page 9.

Adult. Gannet. Immature.

Booby Gannet.

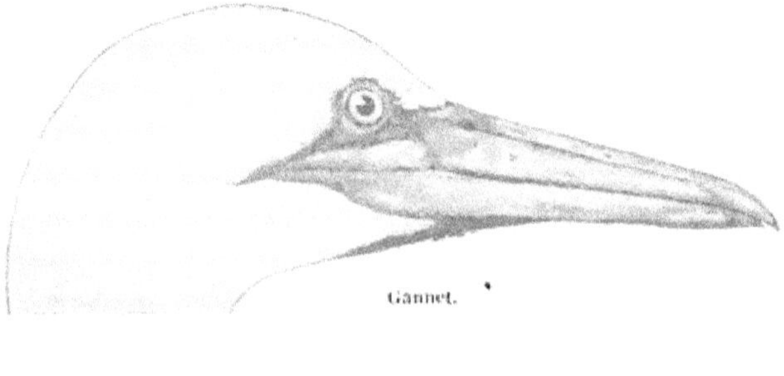

Gannet.

FAMILY ANHINGIDÆ.

Darters. Snake Birds.

Bill, sharp-pointed and slender; chin, bare; neck, long and slender; toes, four, all connected by webs.

* Group 1. Wing, 12.50 to 15.50 inches long.

Neck, long, snake-like; head and neck, black in male; brown in female; outer webs of two middle tail feathers, "fluted." *Anhinga anhinga.* **Snake Bird. Water Turkey.**

See No. 78.

FAMILY PHALACROCORACIDÆ.

Cormorants.

Bill, hooked at tip, and less than twelve inches long; bare skin at base of bill and chin; lores, bare; toes, four, all connected by webs.

* Group 1. Wing, 11 to 15 inches long.

Tail composed of fourteen feathers.

Phalacrocorax carbo. **Common Cormorant. Shag.**
See No. 79.

Tail composed of twelve feathers.
Phalacrocorax dilophus, and races. **Double-crested Cormorant and Florida Cormorant.**
See Nos. 80 and 81

The Mexican Cormorant, a smaller species, occasionally occurs in the Mississippi Valley.

* For directions for measurement, see page 9.

FAMILY PHALACROCORACIDÆ. — CORMORANTS.

Common Cormorant.

Double-crested Cormorant. Mexican Cormorant.

FAMILY PELECANIDÆ.

Pelicans.

Bill, hooked at tip, over twelve inches long and having a large pouch; lores, bare; toes, four, all connected by webs.

White Pelican.

* Group 1. Wing, over 17 inches long.

General plumage, white. *Pelecanus erythrorhynchos.* **White Pelican.** See No. 83.

General plumage, not white (coloration very variable according to age and season, but never white). *Pelecanus fuscus.* **Brown Pelican.** See No. 84.

* For directions for measurement, see page 9.

Brown Pelican.

General plumage, variable, according to age and season, but *never white*.

Pelecanus fuscus. **Brown Pelican.**

See No. 84.

FAMILY FREGATIDÆ.

Man-of-war Birds. Frigate Birds.

Bill, hooked at tip; lores, feathered; upper plumage, entirely black; toes, four, all connected by webs; tail, forked; wings, very long.

* Group 1. Wing, over 21 inches long.

Entire plumage, black (*adult, male*); general plumage, black; belly, white (*female*); head and neck, whitish; belly, white; rest of plumage, black (*immature*).

Fregata aquila. **Man-of-war Bird.**
See No. 85.

---

* For directions for measurement, see page 9.

FAMILY ANATIDÆ.

DUCKS, GEESE, AND SWANS.

SUBFAMILY MERGINÆ. MERGANSERS.

FISH-EATING DUCKS HAVING NARROW BILLS WITH TOOTH-LIKE SERRATIONS ON EDGES; TARSUS, SCUTELLATE IN FRONT (TRANSVERSE SCALES).

GROUP 1. Wing, 6.50 to 8.50 inches long. See page 61.
GROUP 2. Wing, 8.50 to 12 inches long. See page 63.

SUBFAMILY ANATINÆ. RIVER AND POND DUCKS.

DUCKS HAVING HIND TOE WITHOUT WELL DEVELOPED, MEMBRANEOUS LOBE OR FLAP; TARSUS, SCUTELLATE (TRANSVERSE SCALES) IN FRONT.

GROUP 1. Wing, 5 to 7.50 inches long. See page 63.
GROUP 2. Wing, 8 to 10 inches long. See page 64.
Section 1. Belly, white. See page 64.
Section 2. Belly, not white. See page 64.
GROUP 3. Wing, 10 to 14 inches long. See page 65.
Section 1. Belly, white, or tinged with dusky or gray on lower part. See page 65.
Section 2. Belly, not white. See page 66.

SUBFAMILY FULIGULINÆ. BAY AND SEA DUCKS.

DUCKS HAVING A FLAP OR MEMBRANEOUS LOBE ON HIND TOE; TARSUS, SCUTELLATE (TRANSVERSE SCALES) IN FRONT.

GROUP 1. Wing, 5 to 6.50 inches long. See page 71.
Section 1. Tail feathers, not stiff and pointed. See page 71.
Section 2. Tail feathers, stiff and pointed. See page 71.
GROUP 2. Wing, 6.50 to 7.50 inches long. See page 71.
Section 1. Belly, white. See page 71.
Section 2. Belly, not white. See page 72.
GROUP 3. Wing, 7.50 to 8.50 inches long. See page 72.
Section 1. Head, with more or less white or brownish white; belly, white, sometimes tinged with dusky or gray on lower part. See page 72.
Section 2. No white or grayish white on head; belly, white; sometimes tinged with grayish white on head. See page 73.
Section 3. Belly, not white. See page 73.
GROUP 4. Wing, 8.50 to 10 inches long. See page 74.
Section 1. Hind toe, with flap or lobe; belly, white, sometimes tinged on lower part with dusky or gray; *head, marked with more or less white*, or brownish white. See page 74.

Section 2. Belly, white, sometimes tinged on lower part with gray or dusky; no white or grayish white on head. See page 74.

Section 3. Belly, not white; head, marked with more or less white, or grayish white. See page 76.

Section 4. Belly, not white; no white on head. See page 77.

GROUP 5. Wing, 10 to 14 inches long. See page 77.

Section 1. Head, with more or less white, or grayish white; belly, black. See page 77.

Section 2. No white on head; belly, black. See page 79.

Section 3. Head, with more or less white, or grayish white; belly, mottled brown, or grayish brown, or slaty. See page 79.

Section 4. No white on head; belly, mottled brown, or grayish brown. See page 80.

SUBFAMILY ANSERINÆ. GEESE.

LORES, FEATHERED; TARSUS, RETICULATE.

GROUP 1. Wing, 12 to 14 inches long. See page 81.

GROUP 2. Wing, 14 to 20 inches long. See page 81.

Section 1. Head and neck, black or blackish, marked with more or less white; bill and feet, black. See page 81.

Section 2. Head, white, sometimes tinged with brownish orange; bill and feet, pink or flesh color in life; yellowish or pale brownish in dried skin. See page 83.

Section 3. Head, brownish or grayish, sometimes marked with white; bill, pinkish; feet, yellow or pink. See page 83.

SUBFAMILY CYGNINÆ. SWANS.

BARE SKIN BETWEEN THE BILL AND EYE; TARSUS, RETICULATE; NECK, VERY LONG; WING, OVER 19 INCHES LONG. See page 84.

ORDER ANSERES.

Lamellirostral Swimmers.

FAMILY ANATIDÆ.

Ducks, Geese, and Swans.

SUBFAMILY MERGINÆ.

Mergansers.

Fish-eating Ducks having narrow bills with tooth-like serrations on edges, and the tarsus, scutellate in front.

*Group I. Wing, from 6.50 to 8.50 inches long.

Section I. Bill, narrow, with tooth-like serrations.

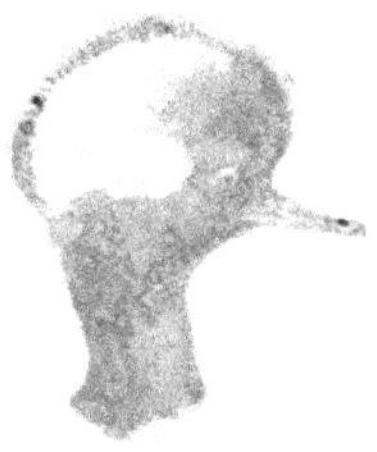

[illegible]

[illegible]

Head, brownish or grayish; tarsus less than 1.50; crest, [illegible] [illegible] **Hooded Merganser.** [illegible]

Head, with black and white crest; [illegible] **Hooded Merganser.** [illegible]

* [illegible]

Red-breasted Merganser.

Male. American Merganser. Female.

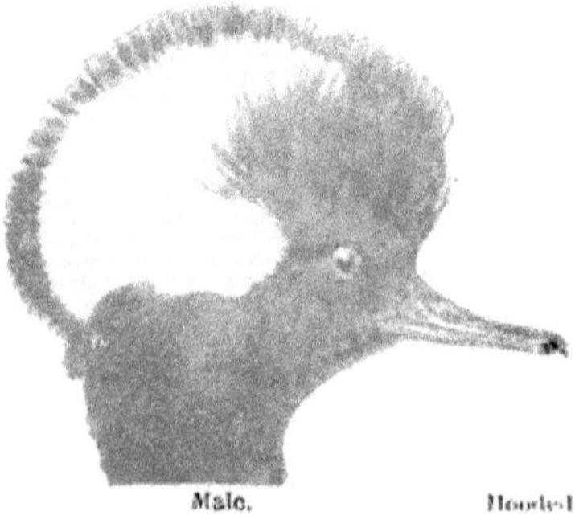

Male. Hooded Merganser. Female.

*Group 2. Wing, from 8.50 to 12 inches long.

Section 1. Bill, narrow, with tooth-like serrations.

Distance from nostril to tip of bill, **less than 1.50**; head and neck, **greenish black**; underparts, creamy white, tinged with salmon color; feet, red (male).
Merganser americanus. **American Merganser. Sheldrake.**
See No. 86.

Top of bill.
Merganser americana.

Distance from nostril to tip of bill, **less than 1.50; head, rufous brown**; upper throat, white; feet, orange red (female).
Merganser americanus. **American Merganser. Sheldrake.**
See No. 86.

Distance from nostril to tip of bill, **more than 1.50; head, black, tinged with green**; breast, rufous, streaked with black (male). *Merganser serrator.* **Red-breasted Merganser.**
See No. 87.

Merganser serrator.

Distance from nostril to tip of bill, more than 1.50; head, brownish, palest on the throat; speculum, white (female). *Merganser serrator.* **Red-breasted Merganser.**
See No. 87.

SUBFAMILY ANATINÆ.

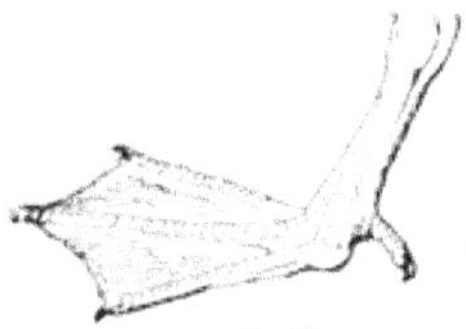

River and Pond Ducks.

Ducks having hind toe without membraneous lobe or flap; tarsus, scutellate in front.

* Group 1. Wing, 5 to 7.50 inches long.

Section 1. Hind toe, without flap or lobe.

Anas discors.

Large patch pale blue on wing (coverts); **white crescent on face** (male); no white crescent, face speckled; **chin whitish** (female). Common in Eastern United States.

Anas discors. **Blue-winged Teal.**
See No. 97.

Large patch of pale blue on wing (coverts); head and breast, rufous brown; crown,

* For directions for measurement, see page 9.

blackish (male): **sides of head, speckled** (dull white, dotted with black); **chin and throat, dusky, tinted with rufous** (female). Western species rare east of the Mississippi River.

Anas cyanoptera. **Cinnamon Teal.**
See No. 98.

No blue patch on wing; head, rufous brown with large patch of green through eye to nape; speculum, black and green (male); head, speckled. No blue patch on wing (female).

Anas carolinensis. **Green-winged Teal.**
See No. 96.

* Group 2. Wing, from 8 to 10 inches long.

Section 1. Belly, white; no flap or lobe on hind toe.

Head, green, purple, black, and white; chin and upper throat, white; feet, yellow in life (male).

Aix sponsa. **Wood Duck. Summer Duck.**
See No. 101.

Head, grayish brown with white stripe through the eye; chin and upper throat, white; toes, dull yellow in life (female).

Aix sponsa. **Wood Duck. Summer Duck.**
See No. 101.

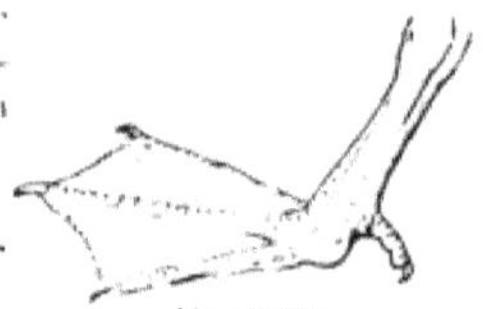

Aix sponsa.

Head, brown; chin, not white; a white stripe on sides of the neck; tail, pointed (male).

Dafila acuta. **Pintail Duck.**
See No. 100.

Section 2. Belly, not white; hind toe, without membraneous lobe or flap.

Spatula clypeata

Bill, more than one inch wide near tip; **head, green or greenish; speculum, metallic green; axillars, white;** feet, orange red in life (male).

Spatula clypeata. **Shoveller Duck. Broad-bill.**
See No. 99.

Bill, more than one inch wide near tip: **head, narrowly streaked and speckled with brown** and dull white; **speculum, metallic green;** feet, *orange red* in life; axillars, white (female).

Spatula clypeata. **Shoveller Duck. Broad-bill.**
See No. 99.

---

* For directions for measurement, see page 9.

Bill, less than one inch wide near tip; **head, lined** and **speckled with brown** and brownish white; speculum, *not metallic green*; **axillars, white, barred with brown**; rump and tail coverts, brown, narrowly edged and banded with white; feet, dusky (female).

Axillar. Dafila acuta.

Dafila acuta. **Pintail Duck.**
See No. 100.

Belly, white, **with more or less brown spots; axillars, white;** bill, less than .90 wide at widest part (female).

Anas strepera. **Gadwall.**
See No. 92.

Belly, more or less spotted; throat, white; rump, olive brown; secondaries, metallic green, tipped with white; axillars and under wing coverts, heavily barred; toes, yellowish in life (female).

Aix sponsa. **Wood Duck. Summer Duck.**
See No. 101.

*Group 3. Wing, from 10 to 14 inches long.

Section 1. Hind toe, without membraneous lobe or flap; belly, white, sometimes faintly tinged with dusky or gray on lower part.

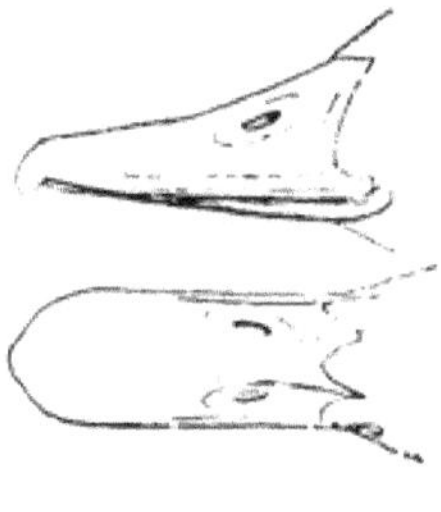

Axillars, white; the shafts, white; exposed speculum, black and white; **head, tawny brown; cheeks and throat, tawny, speckled with brown** (male).

Anas strepera. **Gadwall. Creek Duck.**
See No. 92.

Axillars, white; the shafts, white; head, speckled; exposed speculum, black and white (female).

Anas strepera. **Gadwall. Creek Duck.**
See No. 92.

Axillars, white with dark shafts; speculum, green and black; white patch on shoulder; **top of head, white** (male). *Mareca americana.* **American Widgeon. Baldpate.**
See No. 94.

* For directions for measurement, see page [illegible]

Axillars, white with dark shafts; no white shoulder patch; **head, speckled** (female).
Anas americana. **American Widgeon. Baldpate.**
See No. 94.

Axillars, grayish white; head, brown, not speckled; stripe of white on sides of neck; tail, pointed; middle feathers, long (male).
Dafila acuta. **Pintail Duck.**
See No. 100.

Section 2. Hind toe, without membraneous lobe or flap; belly, not white.

Speculum, bluish purple edged with white; head, green; a white ring around neck; breast, chestnut; belly, grayish white; feet, orange red; axillars, white (male).
Anas boschas. **Mallard Duck.**
See No. 89.

Anas boschas.

Belly, white, showing more or less brown spots; **axillars, white;** bill, less than .90 wide at widest part; speculum, black and white (female).
Anas strepera. **Gadwall.**
See No. 92.

Anas boschas.

Speculum, bluish purple, edged with white and black; greater wing coverts, with broad band of white; **head, tawny brown, streaked with dark brown;** belly, pale buff, mottled with brown; feet, orange red; axillars, white (female).
Anas boschas. **Mallard Duck.**
See No. 89.

Speculum, purplish blue; no white band on greater wing coverts; head and throat, streaked; throat, not buff, no black spot at base of bill; feet, olive (sometimes red); axillars, white; Eastern North America.
Anas obscura. **Black Duck. Dusky Duck.**
See No. 90.

Speculum, purplish blue; no white band on greater wing coverts; head, pale brown; **upper throat, buff, not streaked;** bill, yellowish olive; nail of bill and spot at base, black; feet, pale orange red; axillars, white. Florida species occasionally observed on Gulf coast to Louisiana.
Anas fulvigula. **Florida Black Duck. Florida Dusky Duck.**
See No. 91.

1. American Merganser. Male.
2. American Merganser. Female.
3. Red-breasted Merganser. Male.
4. Red-breasted Merganser. Female.
5. Hooded Merganser. Male.
6. Hooded Merganser. Female.
7. Mallard Duck. Male.
8. Mallard Duck. Female.
9. Black Duck.
10. Florida Duck.
11. Gadwall. Male.
12. Gadwall. Female.
13. European Widgeon.
14. American Widgeon. Male.
15. American Widgeon. Female.

1. GREEN-WINGED TEAL. Male.
2. GREEN-WINGED TEAL. Female.
3. BLUE-WINGED TEAL. Male.
4. BLUE-WINGED TEAL. Female.
5. CINNAMON TEAL.
6. SHOVELLER DUCK. Male.
7. SHOVELLER DUCK. Female.
8. PINTAIL DUCK. Male.
9. PINTAIL DUCK. Female.
10. WOOD DUCK. Male.
11. WOOD DUCK. Female.
12. REDHEAD DUCK. Male.
13. REDHEAD DUCK. Female.
14. CANVAS-BACK DUCK. Male.
15. CANVAS-BACK DUCK. Female.

1. American Scaup Duck, Male.
2. American Scaup Duck, Female.
3. Lesser Scaup Duck.
4. Ring-necked Duck, Male.
5. Ring-necked Duck, Female.
6. American Golden-eye Duck, Male.
7. American Golden-eye Duck, Female.
8. Barrow's Golden-eye Duck.
9. Buffle-head Duck, Male.
10. Buffle-head Duck, Female.
11. Old Squaw, Male.
12. Old Squaw, Female.
13. Harlequin Duck, Male.
14. Harlequin Duck, Female.
15. Steller's Duck.

1. Northern Eider Duck.
2. American Eider Duck. Male.
3. American Eider Duck. Female.
4. King Eider Duck.
5. American Scoter Duck. Male.
6. American Scoter Duck. Female.
7. White-winged Scoter Duck. Male.
8. White-winged Scoter Duck. Female.
9. Surf Scoter Duck. Male.
10. Surf Scoter Duck. Female.
11. Ruddy Duck. Male.
12. Ruddy Duck. Female.
13. Masked Duck. Male.
14. Masked Duck. Female.

SUBFAMILY FULIGULINÆ.

Bay and Sea Ducks.

Ducks having flap or membraneous lobe on hind toe; tarsus, scutellate (transverse scales) in front.

* Group 1. Wing, from 5 to 6.50 inches long.

Section 1. Tail feathers, not stiff and pointed.

Head, greenish purple, with white patch (male); head, grayish brown with white patch (female or immature); no dark stripes on cheek; bill, less than three fourths inch wide.
Charitonetta albeola. **Buffle-head Duck. Dipper Duck.**
See No. 110.

Charitonetta albeola.

Section 2. Tail feathers, stiff and pointed.

Throat, whitish; cheeks, white or whitish; bill, broad, three quarters inch or more wide.
Erismatura jamaicensis. **Ruddy Duck.**
See No.

Front of head, including cheeks, black (adult); head with black stripes on side (female or immature). Tropical species accidental in the United States.
Nomonyx dominicus. **Masked Duck.**
See No. 123.

* Group 2. Wing, from 6.50 to 7.50 inches long.

Section 1. Belly, white.

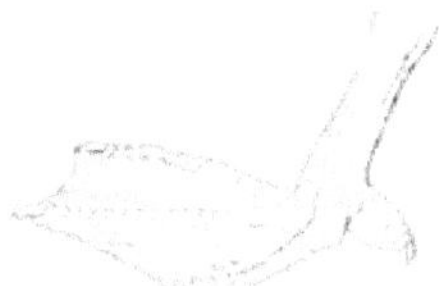

Aythya affinis.

Head, greenish purple, no white on head; back, barred black and white; **speculum, white** (male).
Aythya affinis. **Lesser Scaup Duck. Blue bill.**
See No. 106.

Head, brownish, a patch of dull white at base of bill; no white on ear coverts; **speculum, white** (female).
Aythya affinis. **Lesser Scaup Duck. Blue bill.**
See No. 106.

* [illegible]

Head, greenish purple, no white on head; back, dull black; **speculum, gray** (male).
Aythya collaris. **Ring-neck Duck.**
See No. 107.

Head, dull brown, brownish white at base of bill and below eye; **speculum, gray** (female).
Aythya collaris. **Ring-neck Duck.**
See No. 107.

Head, greenish purple, a large patch of white on back of head (male).
Charitonetta albeola. **Buffle-headed Duck.**
See No. 110.

Head, dark brown or dusky, a **patch of white behind the eye** (on ear coverts); no white or brownish white at base of bill; **speculum, white** (female).
Charitonetta albeola. **Buffle-headed Duck.**
See No. 110.

Section 2. Belly, not white.

Plumage, variously marked with white, slate-color, and chestnut (male).
Histrionicus histrionicus. **Harlequin Duck.**
See No. 112.

General plumage, dull brown, mottled on the under parts (female).
Histrionicus histrionicus. **Harlequin Duck.**
See No. 112.

Head, mostly white, greenish on lores and occiput (male).
Eniconetta stelleri. **Stellers' Duck.**
See No. 114.

Head, light brown; belly, clear, sooty brown (female).
Eniconetta stelleri. **Stellers' Duck.**
See No. 114.

* Group 3. Wing, from 7.50 to 8.50 inches long.

Section 1. Head, marked with more or less white or brownish white; belly, white, sometimes tinged with dusky or gray on lower part.

Head, brownish; a patch of dull white on face at base of bill; **speculum, white** (female).
Aythya marila. **Greater Scaup Duck. Blue-bill.**
See No. 105.

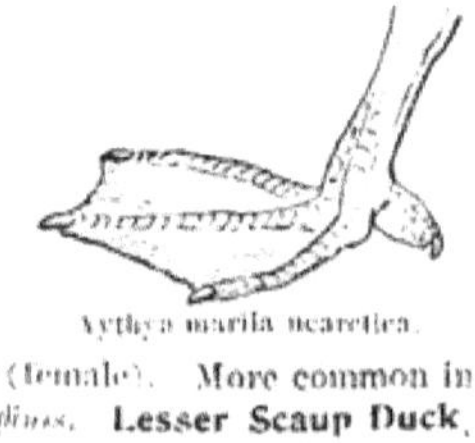

Aythya marila nearctica.

Speculum, *white*, similar to preceding but somewhat smaller (female). More common in the South than the Greater Scaup Duck.
Aythya affinis. **Lesser Scaup Duck.**
See No. 106.

* For directions for measurement, see page 9.

Speculum, gray; head and neck, brownish; chin and anterior portion of lores, brownish white or whitish (female); resembles female Red-head, but is smaller.

Aythya collaris. **Ring-necked Duck.**
See No. 107.

A patch of white or grayish white on the head, including the eye; no speculum; under tail coverts, white; adult males have the tail feathers long and pointed.

Clangula hyemalis. **Old Squaw. Long-tailed Duck.**
See No. 111.

Section 2. Belly, white, sometimes tinged on lower part with dusky or gray; no white or grayish white on head.

Glaucionetta clangula americana.

Head, brown; a distinct wing band of white (female).

Glaucionetta clangula americana. **Golden Eye.**
See No. 108.

Head, black, glossed with green; speculum, white; back, grayish white, finely lined with black (male).

Aythya marila. **Greater Scaup Duck. Blue-bill or Black-head.**
See No. 105.

Head, black, glossed with purple, finely lined with black; **speculum, white;** back, grayish white, smaller than the preceding species, but resembles it closely (male).

Aythya affinis. **Lesser Scaup Duck. Blue-bill or Black-head.**
See No. 106.

Head, black, with violet or bluish gloss, a spot of white on the chin; **speculum, gray;** back, dull black (male). *Aythya collaris.* **Ring-necked Duck.**
See No. 107.

Section 3. Belly, not white.

Plumage, variously marked with white, slate-color, and chestnut; speculum, bluish; top of head and wing coverts, not white (male). *Histrionicus histrionicus.* **Harlequin Duck.**
See No. 112.

Top of head and wing coverts, not white; plumage, dull brown, mottled on the underparts (female). *Histrionicus histrionicus.* **Harlequin Duck.**
See No. 112.

Top of head, white; wing coverts, white; rump, bluish black.

Eniconetta stelleri. **Stellers' Duck.**
See No. 111.

*Group 4. Wing, 8.50 to 10 inches long.

Section 1. Hind toes, with well defined membraneous lobe or flap; belly, white, sometimes tinged on lower part with dusky or gray; head, marked with more or less white, or brownish white.

Aythya marila.

Head, brownish; a patch of dull white on face at base of bill; speculum, white (female).

Aythya marila. **Greater Scaup Duck. Blue-bill. Black-head.**
See No. 105.

Head and neck, dark glossy green; a nearly round patch of white on cheek at base of bill; back, black; speculum, white (male).

Glaucionetta clangula americana. **American Golden Eye. Whistler.**
See No. 108.

Head, bluish black, or purplish blue; an irregular white patch on cheek at base of bill (male).

Glaucionetta islandica. **Barrow's Golden Eye.**
See No. 109.

Top of head, brown; sides of head, brownish white; speculum, gray; bill, more than one and a quarter inches long (female).

Aythya americana. **Red-headed Duck.**
See No. 103.

Bill, not over 1.25 long; **a patch of white, or grayish white, on the head, including the eye;** no speculum; under tail covert, white; adult males have the tail feathers long and pointed.

Clangula hyemalis. **Old Squaw. Long-tailed Duck.**
See No. 111.

Section 2. Hind toe, with well defined membraneous lobe or flap; belly, white, sometimes tinged on lower part with dusky or gray; no white, or grayish white, on head.

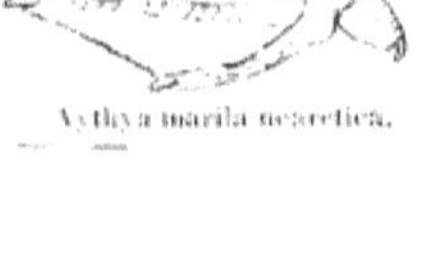

Aythya marila nearctica.

Head, black, glossed with green; back, grayish white, finely lined with black; **speculum, white** (male).

Aythya marila. **Greater Scaup Duck. Blue-bill. Black-head.**
See No. 105

*For directions for measurement, see page 9.

Old Squaws. Male. Female. Harlequin Ducks. Female. Male. Eider Ducks. Male. Female.

Head, cinnamon brown; upper breast and back, ashy gray, not barred; speculum, white (female); very similar to female Barrow's Golden Eye, but slightly smaller, much more common than the next species, on the Atlantic coast.

Glaucionetta clangula americana. **American Golden Eye. Whistler.**
See No. 108.

Head, cinnamon brown; upper breast and back, ashy gray, not barred; speculum, white (female); very similar to preceding species, but slightly larger.

Glaucionetta islandica. **Barrow's Golden Eye.**
See No. 109.

Head, rufous brown; crown, blackish; **breast, black;** lower back, ashy white, finely lined with black (male); bill, very different from that of Red-head. (See cut.)

Aythya vallisneria. **Canvas-back Duck.**
See No. 104.

Aythya vallisneria.

Head, neck, and upper breast, dull cinnamon brown, palest on throat; lower back, dull brown, barred with fine, wavy, white lines (female); easily distinguished by shape of bill. (See cut.)

Aythya vallisneria. **Canvas-back Duck.**
See No. 104.

Aythya americana.

Head, reddish brown; crown, not blackish; breast, black; lower back, grayish, finely lined with black (male); bill, very different from that of Canvas-back. (See cut.)

Aythya americana. **Red-headed Duck.**
See No. 103.

Section 3. Hind toe, with well defined membraneous lobe or flap; belly, not white; more or less white, or grayish white, on head.

General plumage, black; a patch of white on front of crown and nape; bill, large, marked with orange, red, black, and white; no white on wings (male).

Oidemia perspicillata. **Surf Scoter. Skunk-head Coot.**
See No. 121.

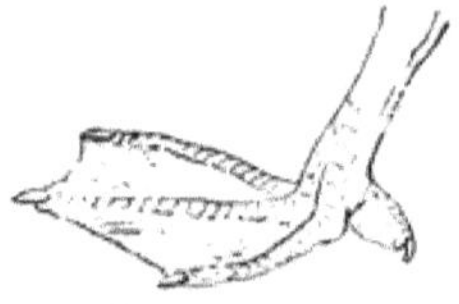

General plumage, brown; top of head and wings, dark; a spot of dull white at base of bill and behind eye; no white on wings (female).

Oidemia perspicillata. **Surf Scoter. Skunk-head Coot.**
See No. 121.

Top and sides of head, white; a patch of greenish in front of eye; throat, black; all **wing coverts, white** (male). *Eniconetta stelleri.* **Stellers' Duck. Stellers' Eider.**
See No. 114.

Section 4. Hind toe, with well defined membraneous lobe or flap; belly, not white; no white on head.

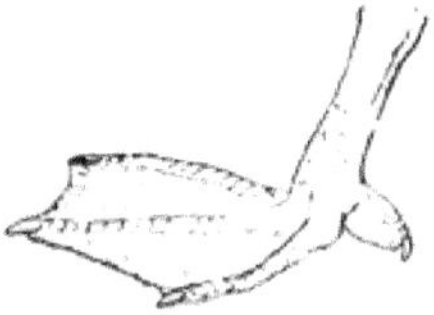

Axillars, brown; secondaries, white, forming a white wing patch; upper parts, dark brown; underparts, sometimes brown, sometimes grayish (female).

Oidemia deglandi. **White-winged Scoter. White-winged Coot.**

See No. 120.

* Group 5. Wing, 10 to 14 inches long.

Section 1. Hind toe, with well defined membraneous lobe or flap; head, with more or less white, or grayish white; belly, black.

General plumage, black; speculum, white; spot under eye, white; axillars, black; bill, orange at base (male).

Oidemia deglandi. **White-winged Scoter. White-winged Coot.**

See No. 120.

Oidemia deglandi.

General plumage, black; head, black, with patch of white on crown and nape; **no white on wing;** axillars, black; bill, orange, black, and white (male).

Oidemia perspicillata. **Surf Scoter. Skunk-head Coot.**

See No. 121.

Dresser's Eider Duck.
Somateria dresseri.

Northern Eider Duck.
Somateria borealis.

Top of head, black, divided on crown; **more or less green on head;** throat, white; axillars, white; bill, culmen, divided and rounded at base (male). See cut of bill.

Somateria dresseri. **American Eider Duck.**

See No. 116.

* For directions for measuring [illegible] page.

Labrador or Pied Ducks.

Camptolaimus labradorius.

Formerly not uncommon on the Atlantic coast. Probably now extinct.

See No. 113.

Top of head, black, divided on crown; **more or less green** on head; throat, white; axillars, white; culmen, divided and pointed at base (male). (See cut of bill.)

Somateria borealis. **Greenland Eider Duck.**
See No. 115.

Top of head, slate color; cheeks, greenish; throat, white, with large, black, **V-shaped mark** (male).

Somateria spectabilis. **King Eider Duck.**
See No. 117.

Section 2. Hind toe, with well defined membraneous lobe or flap; no white on head; belly, black.

General plumage, black; axillars, black; no white on wing; bill, orange at base; feathers on bill, more than one half inch from nostril (male).

Oidemia americana. **American Scoter.**
Black Coot
Butter-bill Coot.
See No. 118.

Section 3. Hind toe, with well defined membraneous lobe or flap; head, marked with more or less white, or grayish white; belly, mottled brown, or grayish brown, or slaty.

General plumage, brownish; no white on wing; **feathers on bill, more than one half inch from nostril** (female and immature).

Oidemia americana. **American Scoter.**
Butter-bill Coot.
See No. 118.

Oidemia americana.

General plumage, grayish brown; speculum, white; feathers on bill, less than one half inch from nostril (female and immature).

Oidemia deglandi. **White-winged Scoter. White-winged Coot.**
See No. 120.

General plumage, grayish brown; feathers extending on upper part of bill more than on the sides; no white on wing; **feathers on bill, less than one half inch from nostril** (female).

Oidemia perspicillata. **Surf Scoter. Skunk-head Coot.**
See No. 121.

Section 4. Hind toe, with well defined membraneous lobe or flap; no white on head; belly, mottled brown, or grayish brown.

General plumage, brownish; no white on wings; **axillars, black; feathers on the bill, more than one half inch from nostril** (female).

Oidemia americana. **American Scoter. Butter-bill Coot. Gray Coot.**
See No. 118.

Head, dark brown or black; **feathers on bill, less than one half inch from nostril; axillars, black; no white on wings** (immature male).

Oidemia perspicillata. **Surf Scoter. Skunk-head Coot.**
See No. 121.

A patch of white on the wings; back and upper parts, dark brown; feathers on the base of bill, extending to within one half inch of nostril (female).

Oidemia deglandi. **White-winged Scoter.**
See No. 120.

Head, tawny, streaked with brown; **axillars, white, or grayish white**; throat, streaked; **feathers on bill, within one fourth inch from nostril; decided difference in bill** from next species (female). (See cut.) Eastern North America.

Somateria dresseri. **American Eider Duck.**
See No. 116.

Head, tawny, streaked with brown; axillars, **white, or grayish white**; throat, streaked; feathers on bill, within **one fourth inch from nostril; decided difference in bill** from preceding species (female). (See cut.) Eastern North America.

Somateria borealis. **Northern Eider Duck.**
See No. 115.

Head, tawny, streaked with brown; axillars, **white, or grayish white**; throat, not streaked; **feathers on bill, more than one fourth inch from nostril** (female).

Somateria spectabilis. **King Eider Duck.**
See No. 117.

SUBFAMILY ANSERINÆ.

Geese.

Lores, feathered; tarsus, reticulate.

*Group 1. Wing, 12 to 14 inches long.

Bill and feet black; **head and neck, black; sides of neck (not front), mottled with white;** no speculum; **lower breast, grayish;** no white on head.

Branta bernicla. **Brant.**
See No. 132.

Head and neck, black; front and sides of neck, mottled with white; no speculum; **lower breast, dark, not pale gray;** no white on head.

Branta nigricans. **Black Brant.**
See No. 133.

A patch of white on side of head, extending to throat; rest of head and **neck, black.** Western United States, rarely to Wisconsin. *Branta canadensis minima.* **Cackling Goose.**
See No. 131.

*Group 2. Wing, from 14 to 20 inches long.

Section 1. Head and neck, black or blackish, marked with more or less white; bill and feet, black.

Head and neck, black; a patch of white on each cheek, extending to upper throat; no white on neck. *Branta canadensis and races.* **Canada Goose.**
See Nos. 129, 130.

Head and neck, black; side *and front* **of neck, speckled with white;** upper belly, whitish. Common on Atlantic coast. *[illegible]* **Brant.**
See No. [illegible]

Head and neck, black; side and front of neck, speckled with white; upper belly, grayish brown. Rare on Atlantic coast. *[illegible]* **Black Brant.**
See No. [illegible]

Face and upper breast, white; lores, black; back of head [illegible]. European species occasionally found on our coast. *[illegible]* **Barnacle Goose.**
See No. [illegible]

* For directions for measurement, [illegible]

Snow Goose.

Brant.

Black Brant.

White-fronted Goose.

Section 2. Bill and feet, pink or flesh color in life; head, white, sometimes tinged with brownish orange.

Entire plumage, white; primaries, black; smaller than *C. h. nivalis*; bill, over 1.90; tarsus, over 2.80; middle toe, over 2.10 (adult). Chiefly Pacific coast to Mississippi Valley, rare on Atlantic coast.
Chen hyperborea. **Lesser Snow Goose.**
See No. 124.

Entire plumage, white; primaries, black; resembles preceding species but is larger (adult). Eastern North America, south in winter on Atlantic coast to Florida and Cuba.
Chen hyperborea nivalis. **Greater Snow Goose.**
See No. 125.

Back, slaty brown; belly and **rump, gray;** the feathers, not barred; terminal half of tail, not white (adult).
Chen cærulescens. **Blue Goose.**
See No. 126.

Section 3. Bill, pinkish; feet, yellow or pink; head, brownish or grayish, sometimes marked with white.

Forehead and feathers at base of bill, white; nail of bill (unguis), whitish; bill, yellowish in dried skin; breast, grayish, more or less marked or spotted with black (adult).
Anser albifrons gambeli. **American White Fronted Goose.**
See No. 128.

No white on forehead or base of bill; bill, yellowish in dried skin; **nail of bill (unguis), dusky;** rump, slaty brown; wing coverts, edged with white (immature).
Anser albifrons gambeli. **American White Fronted Goose.**
See No. 128.

General plumage, grayish; rump, white; smaller than *C. h. nivalis*; bill, over 1.90; tarsus, over 2.80; middle toe, over 2.10 (immature). Chiefly Pacific coast to Mississippi Valley; rare on Atlantic coast.
Chen hyperborea. **Lesser Snow Goose.**
See No. 124.

General plumage, grayish; rump, white; larger than *hyperborea* (immature). Eastern North America, south in winter to Florida and Cuba.
Chen hyperborea nivalis. **Greater Snow Goose.**
See No. 125.

Head, brownish gray; chin, white; rump, gray; unguis (nail of bill), yellow; wing coverts, grayish, showing very little white on the edge of the feathers (immature).
Chen cærulescens. **Blue Goose.**
See No. 126.

SUBFAMILY CYGNINÆ.

Swans.

Bare skin, between the bill and eye; tarsus, reticulate; neck, very long; wing, over 19 inches long.

General plumage, white; bill, black with yellow spot (adult); distance from tip of bill to nostril, less than distance from nostril to eye (immature birds are gray or brownish gray). Common on some parts of the coast. *Olor columbianus.* **Whistling Swan.**
See No. 136.

General plumage, white; bill, black, showing no yellow; distance from tip of bill to nostril, more than distance from nostril to eye. Chiefly found in the interior of North America; not common on the coast. *Olor buccinator.* **Trumpeter Swan.**
See No. 137.

Olor columbianus.

General plumage, gray, or brownish gray; birds of this description may be the young of either of the above species, the difference in the distance from the bill to the eye being characteristic as in the old birds.

FAMILY PHŒNICOPTERIDÆ.

Flamingoes.

Large, tall birds, usually red or pink; bill, very much bent, with tooth-like serrations on edge.

Wing, over 17 inches long (carpus to tip).

General plumage, red or pink; neck and legs, long; bill, much bent, with toothlike serrations on edge. *P*[illegible] **Flamingo.**
See No. 158.

A few Flamingoes are still found in extreme southern Florida. The species is common in the Bahama Islands.

FAMILY PLATALEIDÆ.

Spoonbills.

Bill, wide and flat at the end; toes, four, all on same level.

Wing, 13 to 17 inches long.

Bill, flat, widened and rounded at tip; general plumage, pink and white.

Ajaja ajaja. **Roseate Spoonbill.**

See No. 139.

The Spoonbill may be readily recognized by its peculiar bill. Although by no means common, it still occurs in some numbers in the swamps of southern Florida.

FAMILY IBIDIDÆ.

Ibises.

Bill, long, rather slender, and decidedly curved downward; tarsus, always less than five inches long; toes, four, all on the same level, no comb-like edge on side of middle toe nail; wing, from 8.50 to 13 inches long.

Wing, less than 10.50 inches long.

General plumage, purplish chestnut, showing purplish reflections on head, and greenish on wings; axillars and under wing coverts, purplish bronze; **feathers, bordering base of bill, whitish.** *Plegadis guarauna.* **White-faced Glossy Ibis.**

General plumage, dark chestnut; **feathers, at base of bill, not white.**
Plegadis autumnalis. **Glossy Ibis.**
See No. 142.

General plumage, white; primaries, blackish (adult). [illegible] **White Ibis.**
See No. [illegible]

General plumage, brownish or grayish, often more or less mixed [illegible] **rump, white** (immature). [illegible] **White Ibis.**
See No. [illegible]

General plumage, scarlet. The young Scarlet Ibis somewhat [illegible] White Ibis, but differs in having the rump not white. **Scarlet Ibis.**
See No. [illegible]

FAMILY CICONIIDÆ.

Storks and Wood Ibises.

Greater part of plumage, white; bill, rounded and somewhat curved, very thick and strong; tarsus, always over 5 inches long; toes, four, all on same level; no comb-like edge on inner side of middle toe nail.

Wing, 17 to 19 inches long.

General plumage, white; the wings and tail, more or less black; adult birds have the head bare. *Tantalus loculator.* **Wood Ibis.**

See No. 111.

FAMILY ARDEIDÆ.

Herons, Egrets, and Bitterns.

Bill, nearly straight and sharply pointed; inner side of middle toe nail, with distinct comb-like edge; toes, four, all on same level. Bitterns, tail with ten feathers. Herons, tail with twelve feathers.

* Group 1. Wing, less than 6 inches long.

Least Bittern.

Underparts and sides of the head and throat, buff white; [illegible] sides of the breast. [illegible] **Least Bittern.**
See No. [illegible]

Underparts and sides of the head and throat, rufous chestnut; [illegible] dull black. [illegible] **Cory's Least Bittern.**
See No. [illegible]

* Group 2. Wing, 6 to 7.50 inches long.

Crown, greenish or greenish black; legs, orange yellow in life.

Ardea virescens. **Green Heron.**

See No. 158.

* Group 3. Wing, 8.50 to 11 inches long.

Greater part of plumage, slaty blue; **belly, slaty blue;** head, tinged with purplish brown (adult).

Ardea caerulea. **Little Blue Heron.**

See No. 157.

* For directions for measurement, see page 9.

General plumage, white; **tips of primaries, tinged with slaty blue**; legs, yellowish olive (immature).

Ardea cœrulea. **Little Blue Heron.**

See No. 157.

General plumage, white; ends of primaries, not tinged with slaty blue; **legs, black**; feet, yellow.

Ardea candidissima. **Snowy Heron.**
Snowy Egret.

See No. 151.

General plumage, **tawny brown,** mottled and streaked with dark brown; upper surface of primaries, **blackish.** *Botaurus lentiginosus.* **American Bittern.**
See No. 145.

Night Heron

Bill, large; **top of head, black;** back, green (adult).
Nycticorax nycticorax naevius.
Night Heron.
See No. 159.

Bill, large; **top of head, white or whitish;** back, not green (adult).
Nycticorax violaceus.
Yellow-crowned Night Heron.
See No. 160.

Bill, large; general plumage, mottled, and streaked brown and white; outer edge of primaries, reddish brown (immature).
Nycticorax nycticorax naevius.
Night Heron.
See No. 159.

Bill, large; general plumage, mottled and streaked brown and white; primaries, slaty brown (immature).
Nycticorax violaceus.
Yellow-crowned Night Heron.
See No. 160.

* Group 4. Wing, 11 to 15 inches long.

Bill, stout; top of the head, black; back, dark green (adult).

Nycticorax nycticorax naevius. **Night Heron.**
See No. 159.

Bill, stout; top of head, white or whitish; back, not green (adult).

Nycticorax violaceus. **Yellow-crowned Night Heron.**
See No. 160.

Bill, stout; general plumage, mottled and streaked brown and white; **outer edge of primaries, reddish brown** (immature). *Nycticorax nycticorax naevius.* **Night Heron.**
See No. 159.

Bill, stout; general plumage, mottled and streaked grayish brown and white; **primaries, slaty brown** (immature). *Nycticorax violaceus.* **Yellow-crowned Night Heron.**
See No. 160.

General plumage, white; **bill, yellow**; legs, black. *Ardea egretta.* **American Egret.**
See No. 153.

General plumage, white; legs, dark olive; **terminal half of bill, black; basal half, flesh color** (white phase). *Ardea rufescens.* **Reddish Egret.**
See No. 155.

Head and neck, rufous chestnut; rest of **plumage, slate color,** or slaty gray.

Ardea rufescens. **Reddish Egret.**
See No. 155.

General plumage, yellow brown, mottled and streaked with dark brown; upper surface of primaries, blackish. *Botaurus lentiginosus.* **American Bittern.**
See No. 145.

* Group 5. Wing, over 15 inches long.

Plumage, entirely white; wing, over 17 inches; bill, over 5.50; tarsus, over 7.25.
Ardea occidentalis. **Great White Heron.**
See No. 148.

Plumage, entirely white; wing, less than 17 inches long; bill, under 5.50; tarsus, under 7.25. *Ardea egretta.* **American Egret.**
See No. 153.

Greater part of upper plumage, bluish gray or slaty gray; adults in breeding have middle of crown and throat, white; bill, less than 6.25; immature birds have the top of the head, black. *Ardea herodias.* **Great Blue Heron.**
See No. 151.

Similar to Great Blue Heron, but larger; bill, over 6.25. Occurs only in Florida.
Ardea wardi. **Ward's Heron.**
See No. 150.

Top of head and sometimes greater portion of head, white; wings and back, bluish gray, or slaty gray; general appearance of Ward's Heron except color of head. Occurs only in Florida; probably a color phase of *Ardea occidentalis.*
Ardea wuerdemanni. **Wuerdeman's Heron.**
See No. 149.

* For directions for measurement, see page 9.

ORDER PALUDECOLÆ.

Cranes, Rails, Courlans.

FAMILY GRUIDÆ.

Cranes.

Bill, over 3 inches long; wing, over 16 inches; tarsus, over 7 inches; toes, four, no comb-like edge on inner side of middle toe nail; hind toe, elevated above level of front toes.

General plumage, white; primaries, black; wing, over 20 inches long (carpus to tip); immature birds are more or less buff. *Grus americana.*

Whooping Crane.
See No. 164.

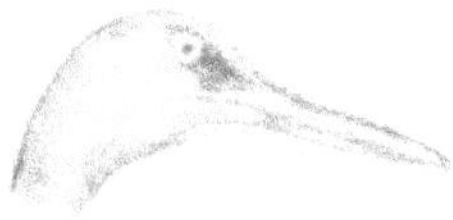

General plumage, slaty gray; bare skin, on head of adult; red, in life; immature birds are brownish; wing, from 15 to 22 inches long. *Grus mexicana.*

Sandhill Crane.
See No. 163.

FAMILY ARAMIDÆ.

Courlans. Limpkins.

Bill, over 3 inches long; tarsus, under 7; wing, under 16; toes, four, no comb-like edge on inner side of middle toe nail; hind toe, raised above level of front toes.

General plumage, dark olive brown, streaked and marked with pure white; tail, purplish brown showing metallic gloss when held in the light; feathers of the back, breast, and wing coverts, brown, with white stripe in middle of each feather; tip of lower mandible often slightly twisted. Occurs in Atlantic States only in Florida. *Aramus giganteus.* **Limpkin.**

See No. 161.

FAMILY RALLIDÆ.

Rails, Gallinules, and Coots.

SUBFAMILY RALLINÆ.

Rails.

Birds which frequent marshy places. Toes, long; wings, short and rounded; bill, shorter than middle toe and claw together.

* Group 1. Wing, less than 3.75 inches long.

Black Rail.

Throat, breast, and **sides of head, slaty gray**; primaries with more or less white spots. Frequents marshes. *[illegible]* **Black Rail.**
See No. 17[illegible].

* Group 2. Wing, from 3.75 to 4.75 inches long.

Virginia Rail.

Back, black or fuscous, the feathers edged with brownish or grayish olive; **underparts, cinnamon rufous,** whitish on the throat; bill, slightly curved.

Rallus virginianus. **Virginia Rail.**

See No. 169.

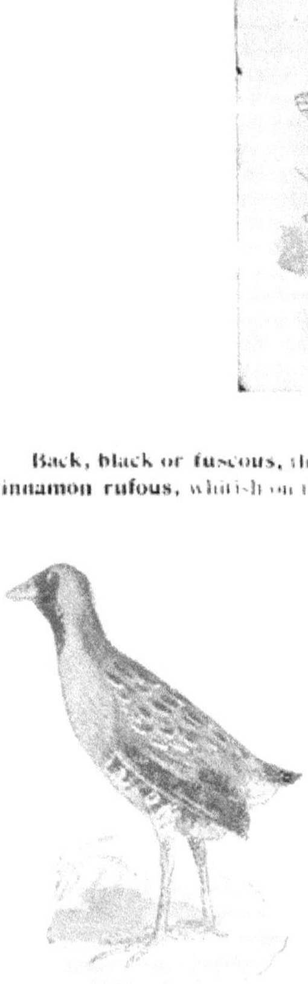

Back, with more or less white streaks; breast, gray or tawny, according to age; lower belly, dull white.

Porzana carolina. **Carolina Rail or Sora.**

See No. 171.

* For directions for measurement, see page 9.

* Group 3. Wing, 4.75 to 6.50 inches long.

Feathers on back, dark olive, edged with gray; cheeks and ear coverts, pale cinnamon rufous, sometimes blackish in young birds; bill, over 1.25 inches long. Usually found near fresh water. *Rallus elegans.* **King Rail.**
See No. 165.

Feathers on the back, black, edged with grayish olive; cheeks and ear coverts, gray. Prefers salt-water marshes. *Rallus longirostris crepitans* and subspecies. **Clapper Rail.**
See No. 166.

* Group 4. Wing, over 6.50 inches long.

Virginia Rail. King Rail. Black Rail. Sora Rail.

Breast, tinged with rufous cheeks; rufous throat, whitish; back, streaked; bill, long and slightly curved.

Rallus elegans. **King Rail.**

See No. 165.

SUBFAMILY FULICINÆ.

Coots.

Bill, short and pointed; forehead, with more or less of a shield; toes, with large lobate webs; head, blackish; back, dark slaty gray; underparts, dark ash gray, whitish, on abdomen.

Fulica americana. **American Coot. Mud Hen. Blue Peter.**

See No. 178.

* For directions for measurement, see page 9.

SUBFAMILY GALLINULINÆ.

Gallinules.

Toes, not webbed; no white on sides of body; head and breast, purplish blue; back, greenish.

Ionornis martinica. **Purple Gallinule.**

See No. 175.

Head, smoky black; breast, dark gray; toes, not webbed; more or less white on sides of body.

Gallinula galeata. **Florida Gallinule.**

See No. 179.

Coot. Gallinule.

ORDER LIMICOLÆ.

SHORE BIRDS.

HIND TOE, WHEN PRESENT, ALWAYS ELEVATED ABOVE LEVEL OF FRONT TOES.

FAMILY PHALAROPIDÆ. PHALAROPES.

SIDES OF TOES WITH ROUNDED LOBES OR NARROW WEBS. See page 104.

FAMILY RECURVIROSTRIDÆ. AVOCETS AND STILTS.

TOES, MORE OR LESS WEBBED (NOT LOBATE); TARSUS, OVER 3.50; BILL CURVED UPWARD OR STRAIGHT. See page 106.

FAMILY SCOLOPACIDÆ.

SNIPE, SANDPIPERS, CURLEWS, WOODCOCK, GODWITS, WILLETS, ETC.

TARSUS, LESS THAN 3.50 INCHES LONG; SIDES OF TOES, WITHOUT LOBATE WEBS; TOES, FOUR (ONE EXCEPTION, THE SANDERLING).

GROUP 1. Birds having wings from 3.25 to 3.75 inches long. See page 107.
GROUP 2. Birds having wings from 3.75 to 4.50 inches long. See page 108.
- Section 1. Toes, four, with small web (not lobate at base). See page 108.
- Section 2. Toes, four, without web. See page 109.
- Section 3. Toes, three; bill, over .75. See page 109.

GROUP 3. Birds having wings from 4.50 to 5.50 inches long. See page 109.
- Section 1. Toes, four, a small web between toes; bill, less than .75. See page 109.
- Section 2. Toes, four, a small web between outer and middle toes; bill, over 1.75. See page 110.
- Section 3. Toes, four, without web; bill, over 2 inches long. See page 110.
- Section 4. Toes, four, without web; bill, over 1.10 and less than 1.90. See page 112.
- Section 5. Toes, four, without web; bill, under 1.10. See page 114.
- Section 6. Toes, three; bill, over .60 inch long. See page 115.

GROUP 4. Birds having wings from 5.50 to 6.75 inches long. See page 116.
- Section 1. Toes, four, a small web between outer and middle toes; bill, over 1.80. See page 116.
- Section 2. Toes, four, a small web between outer and middle toe; bill, under 1.80. See page 116.
- Section 3. Toes, four, without web; bill, over 2 inches long. See page 117.
- Section 4. Toes, four, without web; bill, under 2. See page 118.

GROUP 5. Birds having wings from 6.75 to 9 inches long. See page 119.
- Section 1. Toes, four with *more or less web*; bill, curved upwards or straight; bill, over 2.60 inches. See page 119.

Section 2. Toes four, *without web;* bill, nearly straight; bill, over 2 inches long. See page 120.

Section 3. Toes, four, with small web; bill, slightly curved upward or straight; bill, under 2.60 and over 1.50 inches long. See page 121.

Section 4. Toes, four, with small web; bill, nearly straight; bill, less than 1.50 inches long. See page 121.

Section 5. Toes, four, with small web; *bill, curved downward;* bill, over 2 inches long. See page 122.

GROUP 6. Birds having wings from 9 to 12 inches long. See page 123.

Section 1. Toes, four; bill curved downward. See page 123.

Section 2. Toes, four; bill, curved upward or nearly straight. See page 123.

FAMILY CHARADRIIDÆ. PLOVERS.

TOES, THREE (ONE EXCEPTION, BLACK-BELLIED PLOVER); BILL, COMPARATIVELY SHORT AND THICK.

GROUP 1. Wing, 3.75 to 4.50 inches long; toes, three, no hind toe. See page 124.

Section 1. Bill, under .60 inch long. See page 124.

Section 2. Bill, over .60 inch long. See page 125.

GROUP 2. Wing, 4.50 to 5.50 inches long; toes, three, no hind toe. See page 125.

Section 1. Bill, over .60 inch long. See page 125.

Section 2. Bill, under .60 inch long. See page 125.

GROUP 3. Wing, 5.50 to 6.75 inches long; toes, three, no hind toe. See page 128.

GROUP 4. Wing, over 6.75 inches long. See page 128.

Section 1. Toes, three, no hind toe. See page 128.

Section 2. Toes, four. See page 128.

FAMILY APHRIZIDÆ. TURNSTONES.

TOES, FOUR; LOWER BACK AND RUMP, WHITE WITH BLACK BAND. See page 129.

FAMILY HÆMATOPODIDÆ. OYSTER CATCHERS.

TOES, THREE; BILL, OVER 2.50 INCHES LONG.

Section 1. Toes, three; bill, red. See page 130.

ORDER LIMICOLÆ.

Snipes, Plovers, Sandpipers, Curlews, Phalaropes, etc.

Hind toe, when present, always elevated above level of front toes.

FAMILY PHALAROPIDÆ.

Phalaropes.

Sides of toes, with lobes or narrow webs.

Wilson's Phalarope. Red Phalarope. Winter plumage.

Crymophilus fulicarius.

Back, heavily streaked with black and tawny; **belly, reddish brown,** showing more or less white; **toes, with small lobate web;** wing, 5.20 to 5.50.

Crymophilus fulicarius. **Red Phalarope.**
See No. 179.

Phalaropus lobatus.

Back, grayish, streaked with tawny; belly, white; toes, partly webbed; bill, under 1.05; tarsus, under 1; wing, 4 to 4.50.

Phalaropus lobatus. **Northern Phalarope.**
See No. 180.

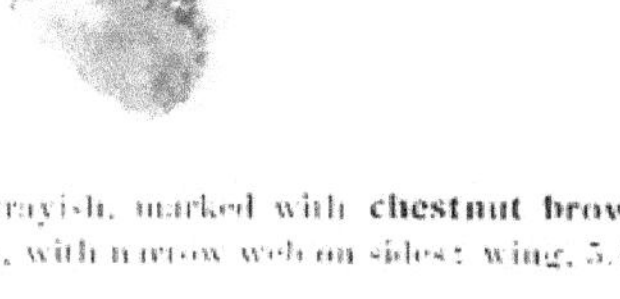

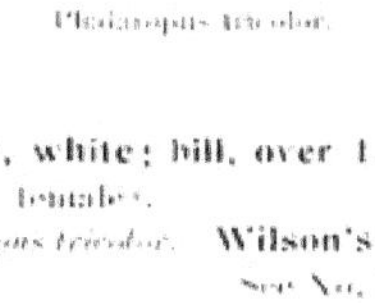

Phalaropus tricolor.

Back, grayish, marked with **chestnut brown; belly, white; bill, over 1.05;** tarsus, over 1; toes, with narrow web on sides; wing, 5.10 to 5.40 (females).

Phalaropus tricolor. **Wilson's Phalarope.**
See No. 181.

Back, grayish, mottled with dusky or whitish; bill, over 1.05; tarsus, over 1 inch; toes, with narrow web on sides; wing, 4.75 to 4.95 (male).

Phalaropus tricolor. **Wilson's Phalarope.**
See No. 182.

FAMILY RECURVIROSTRIDÆ.

Avocets and Stilts.

Tarsus over 3.50 inches; bill, curved upward or straight.

Avocet.

Stilt.

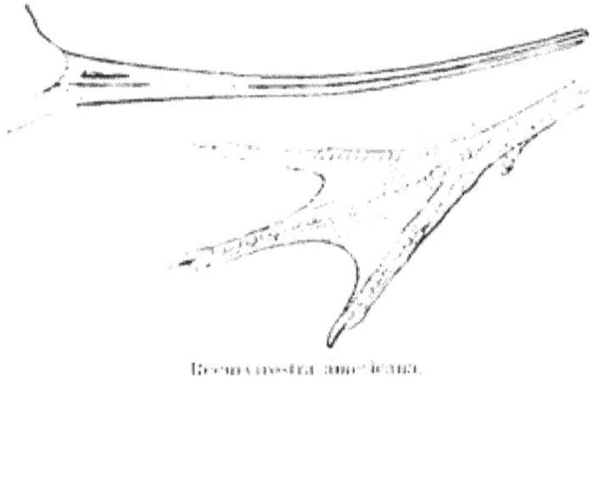

Recurvirostra americana.

Head and neck, pale rufous (summer); white or grayish (winter); back and tail, white; **axillars, white;** belly, white; **first primary, dark with dark shaft;** bill, curved upward; **toes, four.**

Recurvirostra americana.

American Avocet.

See No. 182.

Top of head and nape, black; general upper plumage, glossy black; front of head and front of neck, rump, and underparts, white; **axillars, white; first primary, dark, with dark shaft;** legs, very long, rose pink in life; bill, nearly straight; **toes, three.**

Himantopus mexicanus. **Black-necked Stilt.**
See No. 183.

Himantopus mexicanus

FAMILY SCOLOPACIDÆ.

Snipe, Sandpipers, Curlews, etc.

Tarsus, less than 3.50 inches; toes, four (one exception, Sanderling Sandpiper).

* Group 1. Wing, from 3.25 to 3.75 inches long.

No web between toes; belly, white.

Least Sandpiper.
No. [illegible]

* See [illegible]

Toes, with small web; bill, usually under .85.
Ereunetes pusillus. **Semipalmated Sandpiper.**
See No. 200.

Ereunetes pusillus.
(Foot.)

Toes, with small web; bill, usually over .85.
Ereunetes occidentalis. **Western Sandpiper.**
See No. 201.

* Group 2. Wing, 3.75 to 4.50 inches long.

Section 1. Toes, four, with small web (not lobate) at base.

Ereunetes pusillus.

Bill, under .85; back, not greenish olive; bill, entirely black.
Ereunetes pusillus. **Semipalmated Sandpiper.**
See No. 200.

Bill, over .85; back, not greenish olive; bill, black; no white patch on inner web of third primary. *Ereunetes occidentalis.* **Western Sandpiper.**
See No. 201.

Small web between outer and middle toe; bill, over .85; **back, greenish olive,** sometimes banded; under mandible, pale yellow (in life); third primary and inner primaries with patch of white on inner web. *Actitis macularia.* **Spotted Sandpiper.**
See No. 216.

* For directions for measurement, see page [illegible].

Section 2. Toes, four, without web.

Belly, white; bill, black. *Tringa minutilla.* **Least Sandpiper.**
See No. 196.

[illegible]

Section 3. Toes, three; bill, over .75 inches long.

Belly, white; basal half of outer webs of inner primaries, white; back, mixed rufous, black and white, or grayish brown, or entirely black (breeding), according to season; bill, about one inch long. *Calidris arenaria.* **Sanderling Sandpiper.**
See No. 202.

* Group 3. Wing, 4.50 to 5.50 inches long.

Section 1. Toes, four, a small web between toes; bill, under 1.75.

Bill, under 1.75; tarsus, over 1.30. All other species in this section have the tarsus less than 1.30. *Micropalama himantopus.* **Stilt Sandpiper.**
See No. 190.

Axillars. [illegible]

Tarsus, under 1.30; back, dark olive spotted with white, or brownish gray spotted with dull white, according to season. **Axillars, heavily barred;** a small web between the outer and middle toes. *Totanus solitarius.* **Solitary Sandpiper.**
See No. [illegible]

* For directions for measurement, see [illegible]

Back, greenish olive, *sometimes* barred with black; **axillars, white,** without bars. At some seasons underparts with round black spots; a small web between the outer and middle toe. *Actitis macularia.* **Spotted Sandpiper.**
See No. 216.

Summer. Spotted Sandpipers. Winter.

Section 2. Toes, four, a small web between outer and middle toes; bill, over 1.75.

Macrorhamphus griseus.

Macrorhamphus griseus.

Bill, over 1.75; axillars, white, barred with dark brown; rump and tail, white, spotted and banded with black. *Macrorhamphus griseus.* **Dowitcher. Red-breasted Snipe.**
Macrorhamphus scolopaceus. **Long-billed Dowitcher.**
See Nos. 188 and 189.

Section 3. Toes, four, without web; bill, over 2 inches long.

Axillars, rufous brown, without bars; belly, buff color. *Philohela minor.* **Woodcock.**
See No. 185.

Gallinago delicata.

Axillars, barred black and white; belly, white; upper tail coverts and tail, tawny, more or less marked with black.
Gallinago delicata. **Wilson's Snipe. Jack Snipe.**
See No. 187.

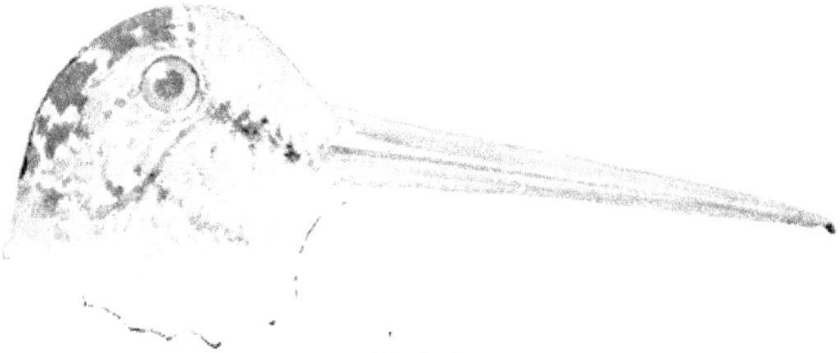

Woodcock.

Wilson's Snipe.

Section 4. Toes, four, without web; bill, over 1.10; and less than 1.90.

Summer. Winter.

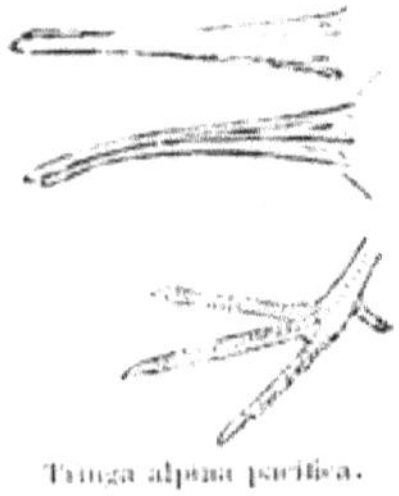

Tringa alpina pacifica.

Bill, decurved near tip; **one or more of inner secondaries, almost entirely white;** *upper tail coverts, not white, barred with black;* **legs and feet, black.** Spring birds have black on the belly, and back, rufous brown and black. Fall birds have the belly white and back gray. *Tringa alpina pacifica.* **Red-backed Sandpiper. American Dunlin.**

See No. 198.

Curlew Sandpiper.

Tringa ferruginea.
Tail and upper tail-coverts.

Bill, decurved near the tip; **upper tail coverts, white, banded with black** or dark brown.

Tringa ferruginea. **Curlew Sandpiper.**
See No. 199.

Pectoral Sandpiper.

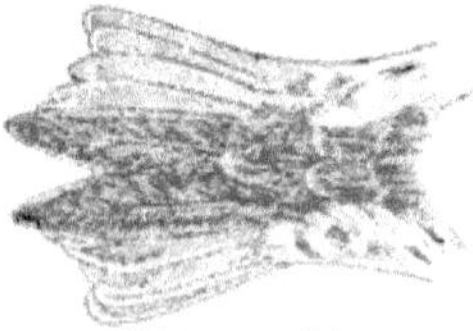

Tringa maculata.
Tail and upper tail coverts.

Bill, nearly straight; back, marked with tawny and black; breast **with numerous narrow, brown streaks;** *none of inner secondaries almost entirely white;* **lower rump and upper tail coverts, black;** the feather more or less tipped with buff; **two middle tail feathers longer than the others.**

Tringa maculata. **Pectoral Sandpiper.**
Grass Bird.

Tringa maculata.

See No. 193.

Bill, nearly straight: **back, dark;** feathers, edged with ashy or buff; **breast, grayish,** without brown streaks; one or more of inner secondaries almost entirely white; **legs and feet, yellow** in life, pale brown in dried skin.

Tringa maritima. **Purple Sandpiper.**
See No. 192.

Purple Sandpiper.

Section 5. Toes, four, without web; bill, under 1.10.

Tringa fuscicollis.

Upper tail coverts, white; inner webs of primaries, not speckled.

Tringa fuscicollis.
White-rumped Sandpiper.
See No. 194.

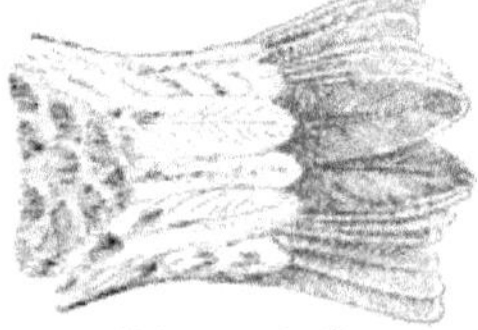

Tringa fuscicollis.

Middle upper tail coverts, smoky or dusky, often tipped with buff; inner webs of primaries not speckled; sides, white; **middle toe and claw, less than .95;** legs and bill, blackish. *Tringa bairdii.* **Baird's Sandpiper.**
See No. 195.

Tringa bairdii.
Tail and upper tail coverts.

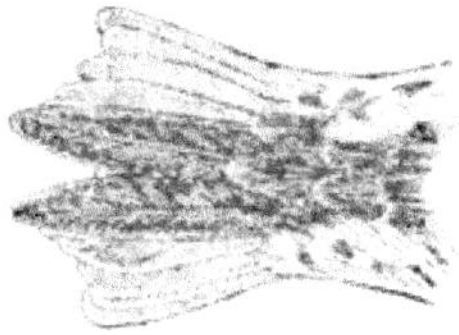

Tringa maculata.
Tail and upper tail coverts.

Middle upper tail coverts, black, often narrowly tipped with brownish buff; inner webs of primaries, not speckled; middle toe and claw, over .95; middle tail feathers decidedly longer than the rest; legs, yellowish olive; base of bill, dull olive; tip, black.
Tringa maculata. **Pectoral Sandpiper. Grass Bird.**
See No. 193.

Inner web of primaries speckled.
Tryngites subruficollis. **Buff-breasted Sandpiper.**
See No. 215.

Tryngites subruficollis.
First primary.

Buff-breasted Sandpiper.

Section 6. Toes, three; bill, over .60 inch.

Calidris arenaria.

Bill, black; shoulder (lesser and middle wing coverts), brown; no web between toes; legs, black. *Calidris arenaria.* **Sanderling Sandpiper.**
See No. 202.

* Group 4. Wing, 5.50 to 6.75 inches long.

Section 1. Toes, four, a small web between outer and middle toe; bill, over 1.80.

Macrorhamphus griseus.

Upper tail coverts and axillars, white, spotted or barred with dusky; bill, nearly straight.

Macrorhamphus griseus. **Red-breasted Snipe, or Dowitcher.**
See No. 188.

Macrorhamphus scolopaceus. **Western Red-breasted Snipe, or Long-billed Dowitcher.**
See No. 189.

Section 2. Toes, four, a small web between outer and middle toe; bill, under 1.80.

Totanus flavipes.

Tarsus and middle toe together, more than 2.60 inches long; outer primary, slate brown, without bars; rump and upper tail coverts, white, more or less barred with brown; **legs, yellow.**

Totanus flavipes. **Summer Yellow-leg.**
See No. 208.

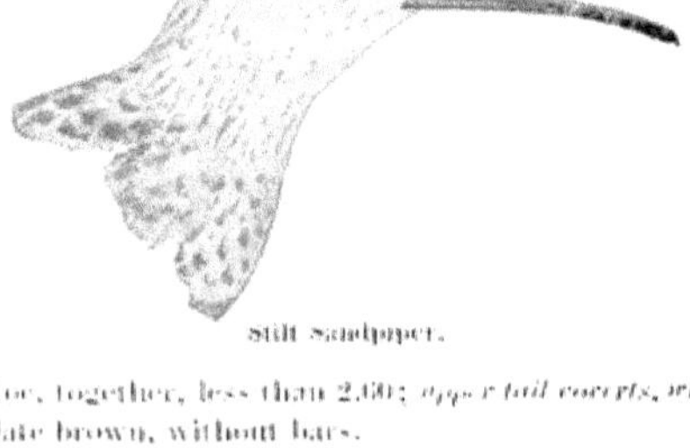

Stilt Sandpiper.

Tarsus and middle toe, together, less than 2.60; *upper tail coverts, white, or white barred with black*; outer primary, slate brown, without bars.

Micropalama himantopus. **Stilt Sandpiper.**
See No. 190.

* For directions for measurement, see page 9.

Upland Plover.

Outer primary, whitish, barred with dark brown.
Bartramia longicauda. **Bartramian Sandpiper. Upland Plover.**
See No. 241.

Section 3. Toes, four, without web; bill, over 2 inches long.

Gallinago delicata.

Lower belly, whitish or white; axillars, barred black and white. *Gallinago delicata.* **Wilson's Snipe. Jack Snipe.**
See No. 187.

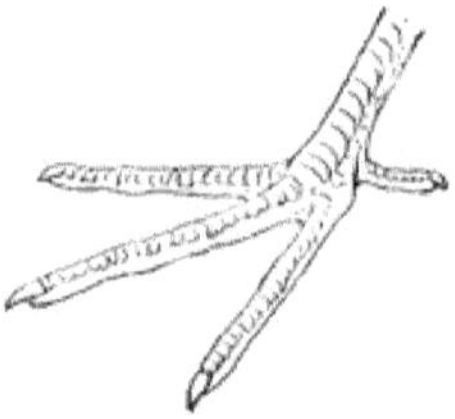

Philohela minor.

Belly, buff color; axillars, rufous brown.
Philohela minor. **Woodcock.**
See No. 185.

Section 4. Toes, four, without web; bill, under 2 inches long.

Rump, gray; upper tail coverts, whitish, banded or marked with black; inner webs of primaries not speckled.
Tringa canutus. **Knot.**
See No. 191.

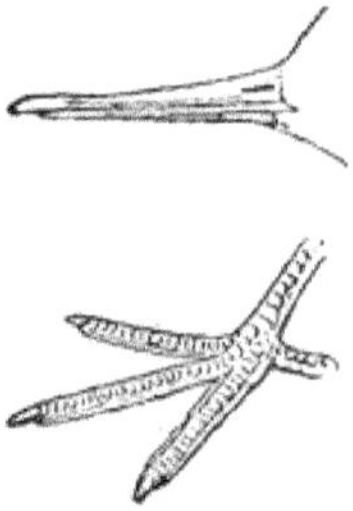

Tringa canutus.

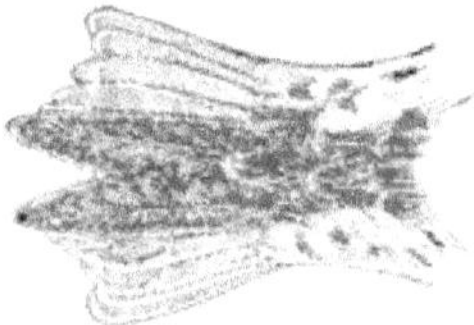

Tringa maculata.

Rump, blackish; middle upper tail coverts, black (not banded); inner web of primaries not speckled.
Tringa maculata. **Pectoral Sandpiper. Grass Bird.**
See No. 193.

Inner webs of primaries, speckled.
Tryngites subruficollis. **Buff-breasted Sandpiper.**
See No. 215.

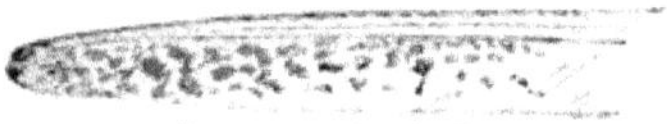

Tryngites subruficollis.

* Group 5. Wing, 6 75 to 9 inches long.

Section 1. Toes, four, with more or less web; bill, curved upwards or straight; bill, over 2.60.

Willet

Symphemia semipalmata.

Axillars, smoky black; belly, white: terminal third of outer primary, black: the rest, white: bill, nearly straight.

Symphemia semipalmata. **Willet.**

Symphemia semipalmata inornata.

Western Willet.

See Nos. 211 and 212.

Axillars, dark gray, or sooty gray: belly, grayish white: **first primary, dark slaty brown with white shaft;** bill, curved upward: **upper tail coverts mostly white.**

Limosa hæmastica.

Hudsonian Godwit.

See No. 201.

Hudsonian Godwit

* [illegible]

Marbled Godwit.

Limosa fedoa.

Axillars, rufous brown; upper tail coverts, not white; belly, buff, sometimes barred with dark brown; **primaries, pale rufous brown with numerous dark dots; shaft of primaries, white;** bill, curved upward.

Limosa fedoa. **Marbled Godwit.**
See No. 203.

Section 2. Toes, four, without web; bill, nearly straight; bill, over 2.60.

Axillars, banded with white and grayish brown; belly, pale brown, banded with dark brown; primaries, grayish brown; **outer web, banded with pale brown or rufous brown;** shaft of primaries, dark; bill, nearly straight.

Scolopax rusticola. **European Woodcock.**
See No. 184.

Section 3. Toes, four (with small web); bill, slightly curved upward or straight; bill, under 2.60 and over 1.50.

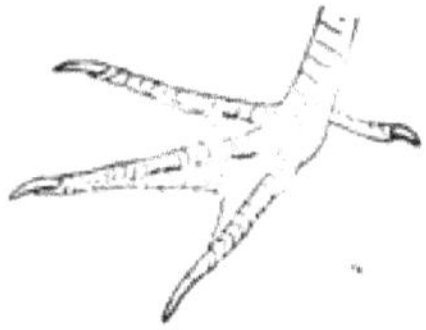

Axillars, smoky black; belly, white; outer primary terminal third, black, rest white; rump, gray; upper tail coverts, white. *Symphemia semipalmata.* **Willet.**

See No. 211.

Axillars, white, with few light brown dots near the ends; belly, white; outer primary dark, with shaft, white; **rump, white without bars; upper tail coverts, white, without bars; legs, olive green.** *Totanus nebularius.* **Greenshank.**

See No. 206.

Axillars, white, banded with brown; belly, white; outer primary, black; shaft, white; **rump, grayish brown; feathers, tipped with white; upper tail coverts, white, more or less barred with dark brown;** legs, **bright yellow.** *Totanus melanoleucus.* **Winter Yellowlegs. Greater Yellowlegs.**

See No. 207.

Section 4. Toes, four, with small web; bill, nearly straight; bill, under 1.50.

Upland Plover.

Axillars, white, banded with brown; [illegible] web, tip, dark. *[illegible]* **Bartramian Sandpiper. Upland Plover.**

See No. [illegible]

Section 5. Toes, four, with small web; bill, curved downward; bill, over 2 inches long.

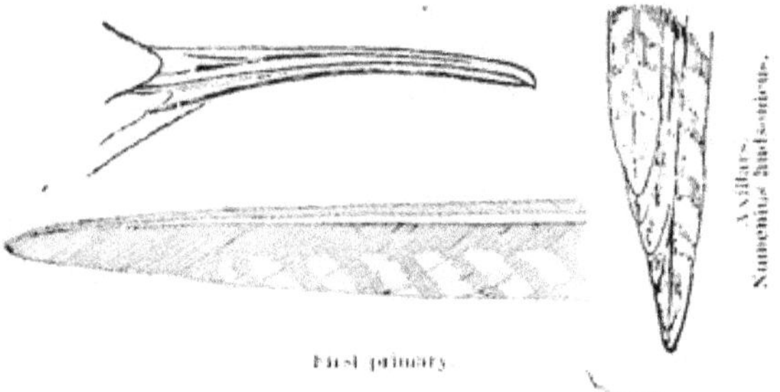

Primaries, barred: axillars barred.

Numenius hudsonicus.
Hudsonian Curlew.
Jack Curlew.
See No. 218.

Primaries, without bars; axillars, barred.

Numenius borealis. **Esquimaux Curlew**
Dough Bird.
See No. 219.

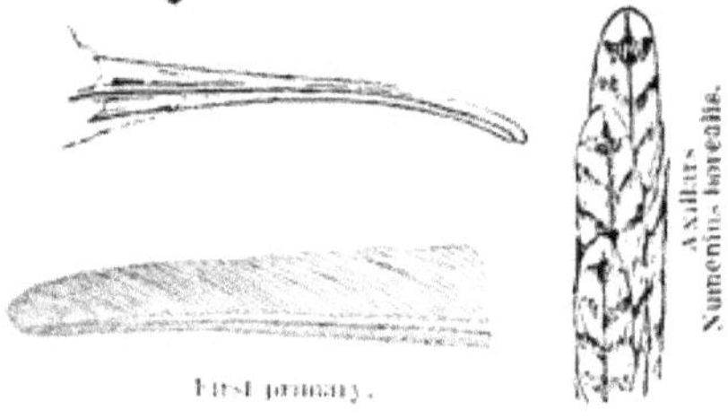

Curlews.

* Group 6. Wing, 9 to 12 inches long.

Section 1. Toes, four; bill, curved downward.

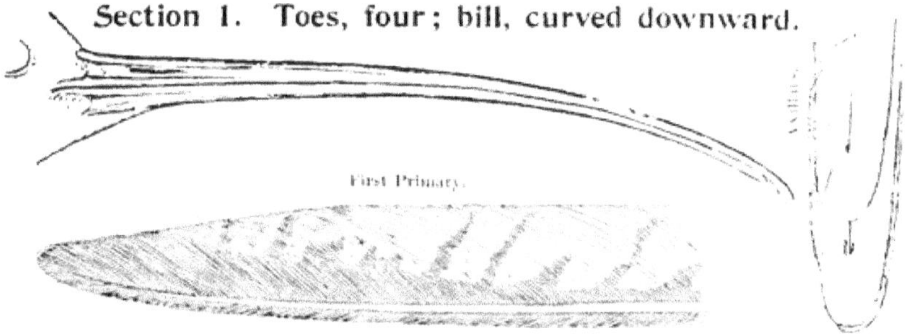

Axillars, reddish brown with narrow black marks; belly, buff; **bill, usually over four inches.** *Numenius longirostris.* **Long-billed Curlew. Sickle-bill Curlew**
See No. 217.

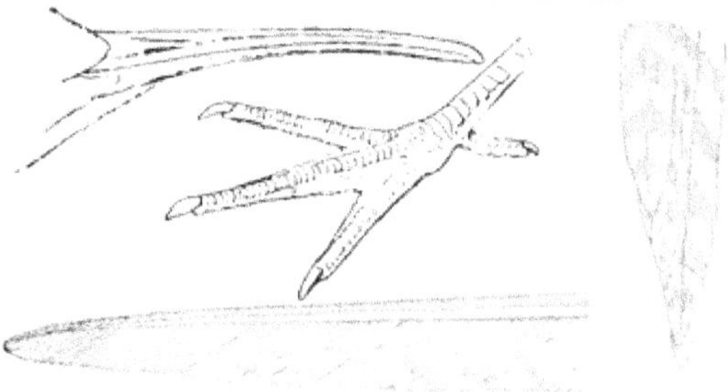

Axillars, banded with slaty brown and dull white; belly, whitish; bill, under four inches.

Jack Curlew.
Hudsonian Curlew.
See No. 218.

Section 2. Toes, four; bill, curved upward, or nearly straight.

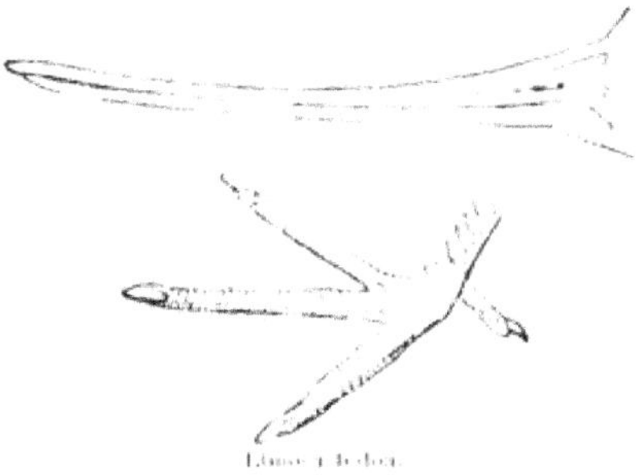

Limosa fedoa.

Axillars, rufous; primaries, rufous, dotted with black.

Limosa fedoa. **Marbled Godwit.**
See No. 2[illegible].

FAMILY CHARADRIIDÆ.

Plovers.

Toes, three, no hind toe.

The Plovers are a cosmopolitan family, numbering something less than one hundred species, fifteen of which occur in North America, including exotic stragglers. As a rule they have but three toes, although two genera, Squatarola and Vanellus, have four. The tarsus is reticulate and the toes are partly webbed.

Black-bellied Plover (Winter). Piping Plover. Semipalmated Plover.

* Group I. Wing, 3.75 to 4.50 inches long.

Section I. Bill, under .60.

Aegialitis semipalmata.

Bill, orange at base, the tip, black; legs, dull flesh color; **a black stripe from bill** passing under eye.

Aegialitis semipalmata. **Semipalmated Plover.**

See No. 226.

* For directions for measurement, see page 9.

Bill, orange at base, the tip, black; legs, orange yellow; two middle tail feathers, tipped with white; **no black stripe from bill to eye;** black breast band not confluent. **Species not found west of the Rocky Mountains.** *Aegialitis meloda.* **Piping Plover.**

See No. 228.

Bill, orange at base, the tip, black; legs, orange yellow; middle tail feathers, tipped with white; no black stripe from bill to eye; **a continuous black band on breast. Species not found west of Rocky Mountains.**

Aegialitis meloda circumcincta. **Belted Piping Plover.**

See No. 229.

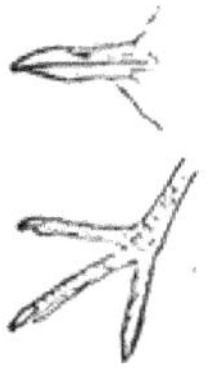

Aegialitis nivosa

Bill, entirely black; legs, slate color; two outer tail feathers, entirely white; two middle feathers, **not** tipped with white; no black stripe from bill to eye. Ranges from Texas and Kansas west to the Pacific Ocean; casual in Western Florida and Cuba; **not known to occur on the Atlantic Coast.** *Aegialitis nivosa.* **Snowy Plover.**

See No. 229a.

Section 2. Bill, over .60 inch long.

Aegialitis wilsonia.

A very small web between outer and middle toes; bill, large and thick; a band of black (male) or brown (female) on breast.

Aegialitis wilsonia. **Wilson's Plover.**

See No. 230.

* Group 2. Wing, 4.50 to 5.50 inches long; toes, three, no hind toe.

Section 1. Bill, over .60 inch long.

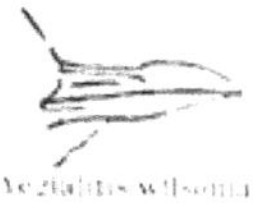

Aegialitis wilsonia

Bill, thick; shoulder (lesser and middle wing coverts), ashy gray; legs, dull flesh color; a small web between toes.

Aegialitis wilsonia. **Wilson's Plover.**

See No. 230.

Section 2. Bill, under .60 inch long.

Aegialitis semipalmata.

Bill, orange at base, the tip, black; legs, dull flesh color; a black stripe from bill passing under eye.

Aegialitis semipalmata. **Semipalmated Plover.** Ring Neck.

See No. 22[illegible].

* For directions [illegible]

Bill, orange at base, the tip, black; legs, orange yellow; no black stripe from bill to eye; black breast band, not confluent; two middle tail feathers, **tipped with white.** Eastern species not found west of Rocky Mountains. *Aegialitis meloda.* **Piping Plover.**
See No. 228.

Bill, orange at base, the tip, black; legs, orange yellow; no black stripe from bill to eye; **breast band, continuous** and not broken in the middle; two middle tail feathers, **tipped with white.** Eastern species not found west of Rocky Mountains.
Aegialitis meloda circumcincta. **Belted Piping Plover.**
See No. 229.

Snowy Plover.

Aegialitis nivosa.

Bill, entirely black; legs, slate color; no black stripe from bill to eye; two middle tail feathers, **not tipped with white;** two outer tail feathers, white. Western species ranges from Texas and Kansas, west, to the Pacific Ocean. Accidental in Florida.
Aegialitis nivosa. **Snowy Plover.**
See No. 229a.

* Group 3. Wing, 5.50 to 6.75 inches long; toes, three, no hind toe.

Breast, with two black bands; underparts, white; **rump and upper tail coverts, orange brown.** *Ægialitis vocifera.* **Killdeer Plover.**
See No. 225.

No black band on breast; back, brownish gray; rump, not orange brown. Western species of casual occurrence in Florida. *Ægialitis montana.* **Mountain Plover.**
See No. 231.

* For directions for measurement, see page 9.

Black-bellied Plover.

Killdeer Plover.

Ring-neck Plover.

Piping Plover.

Wilson's Plover.

Snowy Plover.

* Group 4. Wing, over 6.75 inches long.

Section 1. Toes, three.

Black axillars of Black-bellied Plover.

Gray axillars of Golden Plover.

Charadrius dominicus.

Axillars, gray; rump, not orange brown.
Charadrius dominicus. **Golden Plover.**
See No. 221.

Axillars, smoky black; rump and upper tail coverts, not orange brown.
Charadrius squatarola. **Black-bellied Plover.**
See No. 222.

This species has four toes (the hind toe being so small that it often escapes notice), and properly belongs in Group 5, Section 3, but owing to the fact that it is constantly looked for among the three-toed species it is included in both sections.

Rump, orange brown; underparts, white, with two black bands on the breast; axillars, pure white. *Ægialitis vocifera.* **Killdeer Plover.**
See No. 225.

No black bands on breast; shaft of first primary, white; back, brownish gray; axillars, white. A Western species of casual occurrence in Florida; not recorded elsewhere on Atlantic coast. *Ægialitis montana.* **Mountain Plover.**
See No. 231.

Section 2. Toes, four.

Axillars, smoky black; tarsus, over 1.70; hind toe, very small.
Charadrius squatarola. **Black-bellied Plover.**
See No. 222.

FAMILY APHRIZIDÆ.

Surf Birds and Turnstones.

SUBFAMILY ARENARIINÆ.

Turnstones.

Toes, four; lower back and rump, white with black band.

Summer. Turnstone. Winter.

Adult in summer: General upper parts, mottled and variegated with black, white, rufous, and tawny; throat and breast, black and white; rest of underparts, white; tail, with subterminal band of black, tipped with white.

Adult in winter: Above, light, streaked and dashed with dark brown; an imperfect band of dark brown on the jugulum; chin and upper part of the throat, white; sides of breast, like the back; rest of the underparts, white; a distinct white band on the wing; rump, white, but with a broad patch of black on the upper tail coverts; tail, dark brown, the tips and basal half of the inner feathers, and nearly two thirds of the outer feathers, white; legs, reddish orange; bill, black.

Length, 8.65; wing, 5.70; tail, 2.60; tarsus, 1; bill, .95.

Arenaria interpres. **Turnstone.**
Sp. No. 232.

FAMILY HÆMATOPODIDÆ.

Oyster-catchers.

Toes, three; bill, over 2.50 inches long.

Section 1. Toes, three; bill, red.

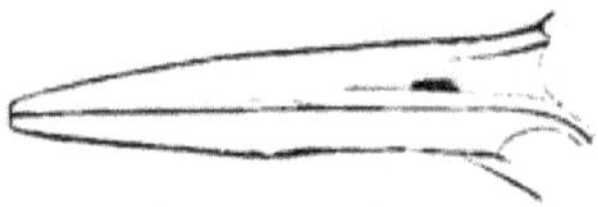

Hæmatopus palliatus.

Head and neck, blackish, or very dark brown; back, brown; upper tail coverts, white; **bill, red;** bill, over 2.50 inches long; wing, about ten inches long; lower breast and belly, white.

Hæmatopus palliatus. **American Oyster-catcher.**

See No. 231.

American Oyster-catcher

ORDER PYGOPODES.

DIVING BIRDS.

SUBORDER PODICIPIDES. GREBES.

FAMILY PODICIPIDÆ. GREBES.

Six species of Grebes occur in North America. Their toes are lobe webbed and the legs are placed far back, rendering walking difficult. They feed principally on fish. About thirty-two species are known throughout the world.

GENUS COLYMBUS LINN.

SUBGENUS COLYMBUS LINN.

Holbœll's Grebe. Pied-billed Grebe.

COLYMBUS HOLBOELLII (*Reinh.*).

Holboell's Grebe.

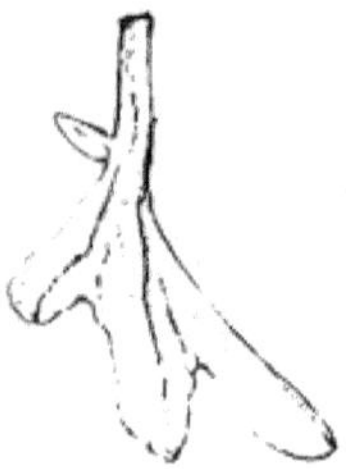

Sp. char. Summer: Crown, nape, and back of the neck, black; back, dull black or blackish; throat, belly, and sides of the head, silvery white; front and sides of neck, rufous brown, gradually shading lighter on breast; sides of body, more or less rufous.

Adult in winter: Upper plumage, sooty brown; throat and underparts, silvery white; neck, more or less tinged with rufous. Immature birds have the throat and sides tinged with gray.

Length, 18 to 20; wing, 7.60; tarsus, 2.45; bill, 1.80 to 2.

Distribution: North America at large, including Greenland (*A. O. U.*); south in winter to North Carolina and Middle States of the interior; breeds from Minnesota and Dakota northwards.

Nest and eggs: The nest is usually composed of grass and reeds, often floating. The eggs are soiled white or pale greenish white, from three to five in number, and measure 2.20 x 1.35.

Holboell's Grebe is the largest of the family in Eastern North America. It is not uncommon on the Atlantic coast during the winter months. Like others of its family it is an expert diver, and rarely takes wing when pursued, usually disappearing beneath the water with an ease and quickness which has won for it the sobriquet of Hell Diver. When frightened it often swims under water with only a small portion of the head and bill exposed.

SUBGENUS DYTES KAUP.

COLYMBUS AURITUS *Linn.*

Horned Grebe. Water Witch.

Adult in summer: Upper parts, including back, wings, top of the head and back of the neck, glossy black; throat, black; front of the neck, breast, and sides of the body, rufous chestnut; rest of underparts, white; wings, dusky black; secondaries, white; lores, dull chestnut; the two small tufts of feathers at the sides of the occiput, brownish buff.

Adult in winter and immature: Head and back, grayish; underparts, white, usually *tinged with gray on the breast* and lower throat; **no tufts** *on sides of the head in winter plumage.*

Length: 13.40; wing, **5.40**; tarsus, **1.70**; bill, .85 to .95.

Distribution: Northern Hemisphere, breeding from the United States northward (*A. O. U.*), migrating south in winter to Florida and the Gulf States.

Nest and eggs: The nest is a mass of floating grass or weeds or a mat of grass on a partly submerged marsh; the eggs are four to seven, dull white or yellowish white, or soiled brownish white, and measure 1.75 x 1.45.

The Horned Grebe is sometimes confounded with the Pied-billed Grebe in winter dress, but the former species has a more pointed and slightly longer bill and tarsus. Both this and the next species are known to gunners by various names, among the most common being, Water Witch, Hell Diver, and Die Dipper.

No. 3. The American Eared Grebe, *C. nigricollis californicus*, has been taken in Illinois. (*Ridgway.*)

GENUS PODILYMBUS LESSON.

PODILYMBUS PODICEPS (LINN.). 4

Pied-billed Grebe. Hell Diver.

Sp. char. Adult in summer: Above, glossy dark brown or brownish black; throat, black, and a black band on the bill; front of neck and sides of neck and body, tinged with pale brown; belly, silvery white.

Adult in winter: Similar, but lacking the black throat and without the black band on the bill; lower belly, tinged with gray.

Length: 13.50; wing, 5; tarsus, 1.45; bill, .90.

Distribution: British Provinces, southward to Brazil, Argentine Republic, and Chili, including West Indies and Bermuda, breeding nearly throughout its range (*A. O. U.*); winters from New Jersey southward.

Nest and eggs: The nest is composed of a mass of floating vegetation or a mat of grass on slightly submerged marshes. The eggs are yellowish white or dirty white, usually from three to eight, and measure about 1.70 x 1.20.

Genus GAVIA. Forster.

GAVIA IMBER (Gunn).

Loon.

Common Loon. Big Loon. Northern Diver.

Adult in summer: Head and neck black, showing green in some lights; a patch of white streaked with black on the throat and sides of the neck; underparts, white; back and wings, black, streaked and spotted with white; where the white spots occur there are usually two spots near the end of each feather; sides of breast, streaked with black.

Adult in winter and immature: Head, grayish; back, grayish brown or dusky brown, without spots; underparts, white; throat, often tinged with gray; primaries, dark; tail feathers, tipped with gray.

Length: 30 to 36; wing, 12 to 14.50; bill, 2.60 to 3.40; height of bill at nostril, about .80; tarsus, 3 to 3.60.

Distribution: Northern portion of Northern Hemisphere, ranging south in winter to Gulf of Mexico and Lower California. Breeds from Northern United States northward.

Nest and eggs: Nest, a mass of grass or weeds and leaves, close to water, more commonly on islets or shores of some pond or lake. Two eggs, grayish brown or greenish brown, heavily blotched with dark brown, and measure 3.50 x 2.25.

Although less numerous than formerly, Loons are common on the Atlantic coast during migrations, and the wild, laugh-like note is a well-known sound on shore and lake.

Summer. Loons. Winter.

INDEX.

A.

B.

M.

N.

O.

P.

INDEX.

U.

W.

X.

Y.

www.ingramcontent.com/pod-product-compliance
Lightning Source LLC
Chambersburg PA
CBHW021658210326
41599CB00013B/1457
* 9 7 8 3 3 3 7 2 2 1 8 3 6 *